AF602154

Indian Tachinid Flies
Biocontrol Agents

The Authors

Dr. Sathe Tukaram Vithalrao [M.Sc., Ph.D., Sangit Vishard, IBT (Seri.), F.I.S.E.C., F.S.E.Sc., F.S.L.Sc., F.I.C.C.B., F.S.S.I., FHAs] is presently working as Professor and Former Head, Department of Zoology, Shivaji University, Kolhapur. He has teaching experience of 32 years in Entomology at University PG department and 21 years in Agrochemicals and Pest Management. He has written 40 books and published 355 research papers in national and international journals of repute. He guided 29 Ph.D. students and completed 6 major research projects (from CSIR, DST, DBT and UGC). He visited Canada (1988), Japan (1988), Thailand (2002, 2004), Spain (2005), France (2005), South Korea (2006) and Nepal (2007) etc. for academic work. He is member of editorial board of 13 prestigious journals. He delivered 35 talks through All India Radio and international conferences and involved in Doordarshan, S.T.V. and B.T.V. programmes on useful and harmful insects. He published more than 35 popular articles in daily newspapers on insects and sericulture. He got several prestigious awards like "Environmentalists of the Year-2003", "Bharat Jyoti", "Jewel of India", "International Gold Star", "Eminent Citizen of India", "Education Acumen", "Best Educationist", "Eminent Scientist of the Year-2008", "Lifetime Education Achievement", "Lifetime Achievement in Zoology (Insect Taxonomy)-2009", Education Leadership-2011, Asia Pacific International Award-2012, Global Education Leadership Award-2013, Indo-Nepal Shiromani Award, etc. He is also working as Research and Recognition (RR) Committee member for Pune University, Pune; North Maharashtra University, Jalgaon; Shivaji University, Kolhapur and DBA Marathwada University, Aurangabad. He has been awarded several fellowships from different scientific and academic societies. He is Chairman of Maharashtra District Environmental Centre of NESA.

Dr. Abhijit S. Desai (M.Sc., Ph.D.) is Research Scientist in Department of Zoology, Shivaji University, Kolhapur. He has published 15 research papers in various journals of national and international repute with impact points 27.02. He is popular and contributory teacher for competitive and graduate courses in Zoology and Life Sciences and attended and presented several research papers in national conferences.

Dr. P.M. Bhoje (M.Sc., Ph.D.), Head, Department of Zoology, Y.C. Warna College, Warnanagar, Kolhapur, Maharashtra has published 25 research papers in various journals of national and international repute. He is BOS member (Zoology) of Shivaji University, Kolhapur and guides 2 Ph.D. students. He visited Thailand and Sri Lanka for presentation of research papers in conferences and President of 'Green Guards' Society, Kolhapur, and NESA district Centre, Kolhapur. He has written 2 books and shared popular chapters in several edited books.

Indian Tachinid Flies
Biocontrol Agents

– Authors –

Prof. (Dr.) T.V. Sathe

Dr. Abhijit S. Desai

Dr. P.M. Bhoje

2017

Daya Publishing House®

A Division of

Astral International Pvt. Ltd.

New Delhi – 110 002

Publisher's Note:

Every possible effort has been made to ensure that the information contained in this book is accurate at the time of going to press, and the publisher and author cannot accept responsibility for any errors or omissions, however caused. No responsibility for loss or damage occasioned to any person acting, or refraining from action, as a result of the material in this publication can be accepted by the editor, the publisher or the author. The Publisher is not associated with any product or vendor mentioned in the book. The contents of this work are intended to further general scientific research, understanding and discussion only. Readers should consult with a specialist where appropriate.

Every effort has been made to trace the owners of copyright material used in this book, if any. The author and the publisher will be grateful for any omission brought to their notice for acknowledgement in the future editions of the book.

Cataloging in Publication Data--DK
Courtesy: D.K. Agencies (P) Ltd. <docinfo@dkagencies.com>

Sathe, T. V., author.
Indian tachinid flies : biocontrol agents / authors, Prof. (Dr.) T.V. Sathe, Dr. Abhijit S. Desai, Dr. P.M. Bhoje.
pages cm
Includes bibliographical references and index.
ISBN 9789389605143 (Int. Edition)

1. Agricultural pests--Biological control--India--Maharashtra. 2. Insect pests--Biological control--India--Maharashtra. 3. Insects as biological pest control agents--India--Maharashtra. 4. Tachinidae--India--Maharashtra. I. Desai, Abhijit S., author. II. Bhoje, P. M. (Prakash Maruti), 1965- author. III. Title.

SB975.5.I4S28 2017 DDC 632.96095479 23

Published by : **Daya Publishing House®**
A Division of
Astral International Pvt. Ltd.
– ISO 9001:2008 Certified Company –
4760-61/23, Ansari Road, Darya Ganj
New Delhi-110 002
Ph. 011-43549197, 23278134
E-mail: info@astralint.com
Website: www.astralint.com

Digitally Printed at : **Replika Press Pvt. Ltd.**

Preface

Tachinids are good biocontrol agents of several types of insect pests including Lepidoptera, Orthoptera, Coleopteran, etc. However, this group of insects is largely neglected from India. Therefore, in the present study 20 species of tachinids were described belonging to four subfamilies, 13 tribes, 15 genera and 1 subgenera from Kolhapur, Sangli, Satara and Sindhudurg districts of Western Maharashtra. 12 new species namely, *W. chandgadensis* sp.n., *P. tricolor* sp.n., *D. indica* sp.n., *C. oriental* sp.n, *C. satarensis* sp.n., *S. kolhapurensis* sp.n., *C. scutellata* sp.n., *A. striatalis* sp.n., *C. abdominalis* sp.n., and *P. panhalensis* sp.n. have been described for the first time and 8 species *T. nigripes, U. uramyiodes, B. schineri, P. siberita* and *D. vacua* etc. have been redescribed. Out of 20 species, 2 species were rare and 15 were common.

Seasonal abundance and distribution of tachinids in the region indicates that, out of four subfamilies Exoristinae was very dominant over other subfamilies. The members of sub-family Exoristinae were abundantly scattered in both agro and forest ecosystems in the region.

Insect pests cause severe destruction to agro and forest crop plants. The tachinids are biopesticides scattered in the region of western Maharashtra. Thus. western Maharashtra is very good natural habitat for tachinid biocontrol agents. They should be identified, conserved, protected and used for pest management as ecofriendly tool. This work will be stimulatory to various workers in the field of pest management and environmental science.

Prof. (Dr.) T.V. Sathe

Dr. Abhijit S. Desai

Dr. P.M. Bhoje

Contents

1
General Introduction

Tachinid flies (Diptera: Tachinidae) are very good biocontrol agents of insect pests economically important both from agro and forest ecosystems. Forest is excellent group of biodiversity and different ecological system including trees with vegetative associations. It is very excellent source medicine, wood, timber, fuel, etc. Forest products are valuable for market and industries. Countries like Russian Federation, Brazil, Canada, United States of America, India, China, Democratic Republic of the Congo, Australia, Indonesia and Sudan together covers 67 per cent of total forest area of the world (Global Forest Resources Assessment, 2010). India is one of the ten most forest rich countries from the world (Global Forest Resources Assessment, 2010). In India (Figure 1), out of 329 million hectares of geographical area only 76.352 million hectares is covered by forest ecosystem. From 413 districts, 105 districts shows forest cover up more than 33 per cent, 252 district between 19 to 33 per cent and 226 districts, less than19 per cent while, 30 districts are devoid of forest covers (FST, 1996). In both, agro and forest ecosystems biodiversity is quite rich in western Maharashtra.

According to Champion and Seth (1968) Indian forests shows four most important groups, *viz.* tropical, subtropical, temperate and alpine which are further divided into 16 subgroups like tropical dry deciduous, tropical moist deciduous, tropical thorn, tropical wet evergreen, subtropical pine, subalpine, Himalayan moist temperate, tropical semi evergreen, montage wet temperate, littoral and swamp, subtropical broad leaved hill, subtropical dry over green, Himalayan dry temperate and sub tropical dry evergreen, *etc.*

Western Ghats is an important feature of Maharashtra. However, it scattered is in Goa, Maharashtra, Tamil Nadu, Karnataka and Kerala states of India (Figure 2). According Myers *et al.* (2000) Western Ghats is one of the world's ten "Hottest biodiversity hotspots" and among the 35 hot spots of the world for biodiversity conservation and protection. The Western Ghats extended from the river Tapi

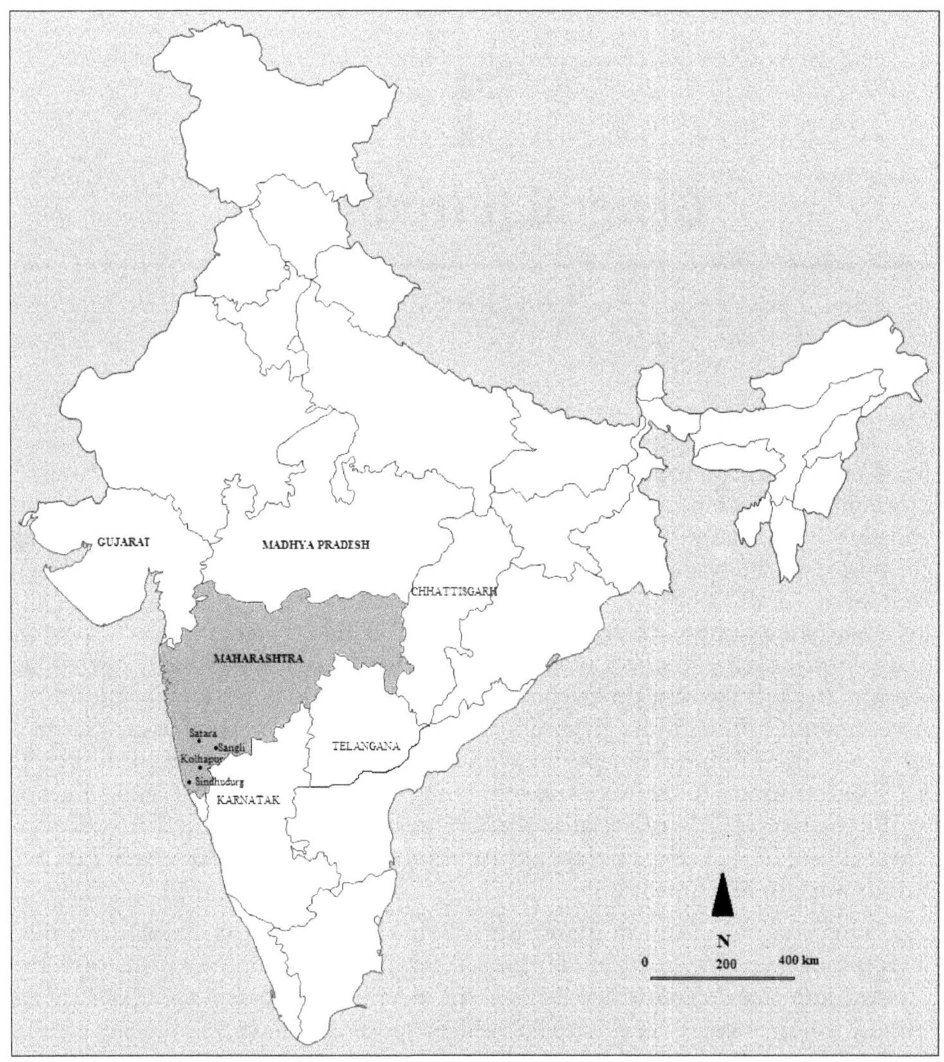

Figure 1: Map of India showing Maharashtra.

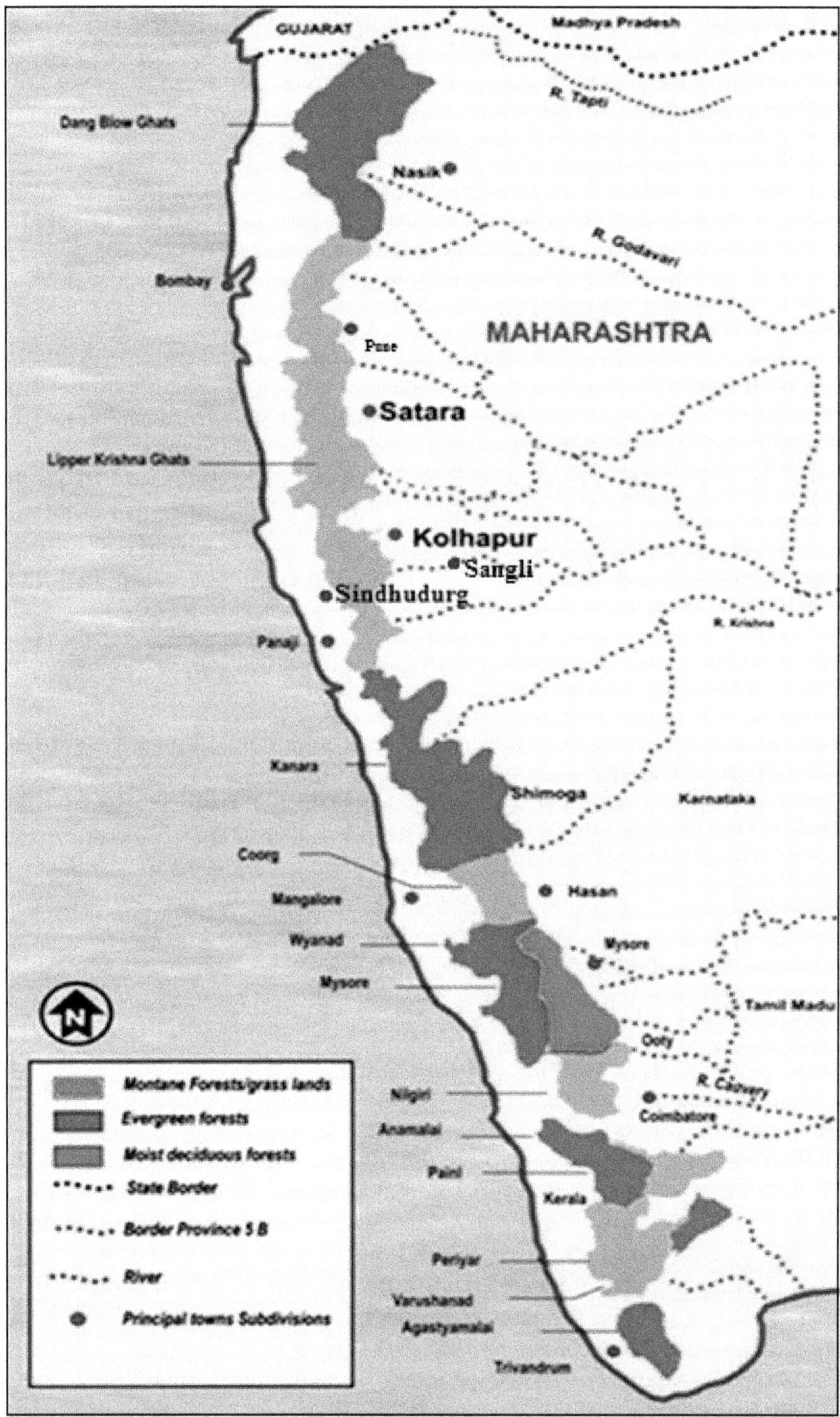

Figure 2: Biographic Subdivision of the Western Ghats.

(21°N) near the border of Gujarat and Maharashtra in the north to Kanyakumari in the south (8°N) (Joseph, 2004). Forest insects include over 16000 species which are nearly one quarter of the 6700 species of insects recorded from different habitats in India (Roonwal *et al.*, 1950).

Insects are abundant and found everywhere in different ecosystems from poles to equator from surface of sea to highest peaks, also from deserts to rain forests. Insects are having very important role in the agroecosystems for crop management (Joseph, 2004; Sathe, 2015). About 6700 species of insects have been described from different ecosystem of India but their behavior, ecology, taxonomy and diversity are still poorly known (ZSI, 1983).

From Shivaji University, Kolhapur Sathe and his several coworkers attempted the taxonomical and ecological work on different groups of insects. Sathe *et al.* (1986; 1987) studied the fauna of butterflies from Western Ghats from Maharashtra and reported 33 butterflies. Out of which 8 species are rare. Gaonkar (1996) studied 330 species of butterflies especially from Western Ghats. Sathe and Pandharbale (1999) described 13 species of Sphingid Moths belonging to six different genera from Western Ghats of Maharashtra. Pandharbale and Sathe (2001) described new species of Syntomid moth, *Syntomis vitex* from forest area of Satara district from Western Ghats of Maharashtra.

Sathe and Shinde (2008) reported 151 species of Damsel and Dragon Flies as biocontrol agents from Koyana. Out of which 17 species of dragon flies are newly introduced to science. Kavane and Sathe (2011) described new 9 subspecies of Wild silk moths *Antheria mylita* from Western regions of Maharashtra. Sathe and Tingare (2007) studied taxonomy of Mosquitoes from different ecosystems of Kolhapur, Satara, Sangli and Solapur districts. Jagtap and Sathe (2011) also studied systematic and molecular phylogeny of Mosquitoes from Western Maharashtra and reported 31 species. Bhusnar and Sathe (2012) studied the diversity of grasshopper from some districts of Western Ghats and reported 54 species, out which 10 species were rare and 4 species newly described. Kadam and Sathe (2015) described 23 species of moths found damaging various plants of Western Ghats.

Although several insects are studied from agro and forest ecosystems by local workers (Sathe, 1992; 2003; 2004; 2012; Sathe *et al.*, 2002; Jadhav and Sathe, 2006; Sathe and Chougale, 2008, Sathe, 2010) very less attention is given on tachinid flies (Diptera) which is well known as biocontrol agents. Today, Diptera is one of the three largest and most diverse animal groups in the world (Skevington and Dang, 2002). True flies are belongs to order Diptera, having near about 100 families. The order Diptera divides into two suborders, Nematocera and Brachycera, the difference between them is depends upon the structure and arrangement of antennae, maxillary palps and larval mandibles. Nematocera have long legs, long antennae and look fragile (*e.g.*, mosquitoes, midges, *etc.*) while Brachycera have stout bodies and short, stout antennae (*e.g.*, house flies, hover flies, tachinid flies *etc.*). About 1, 60,000 species of Diptera have been described in around 10,000 genera and 150 families which cover about 14 per cent of the worlds known insect fauna (Groombridge 1992; Thompson 2005). Dipteran flies are minute to large in length of about 0.5 to 60mm.Variable in form and color, but several of them are soft bodied and flying forms. These insects

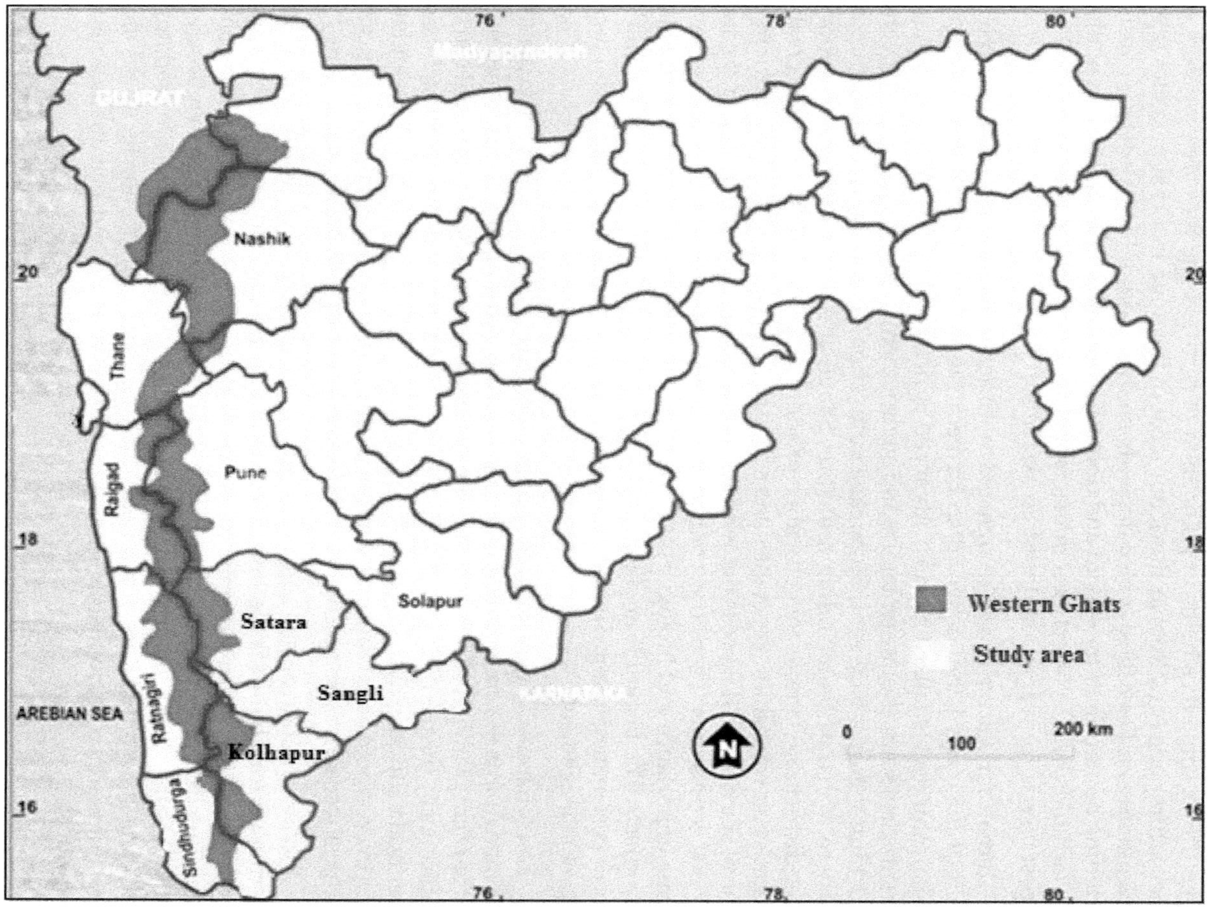

Figure 3: Map of Maharashtra Showing Western Ghats and Study Areas (Kolhapur, Sangli, Satara, Sindhudurg).

are typical because their hind wings are reduced to small, club-shaped structures called as halters which vibrate during flight and help for maintenance their balance.

Some flies like drosophila helps in experimental work like *genetics*, some of the flies are solving crime cases in forensic science (Blow flies, flesh flies) (Jadav and Sathe, 2014), some flies helps in pollination of plants, some flies are predators of other arthropods (*e.g.*, Robber flies), but most of them are external parasites (*e.g.*, Mosquitoes and Deer flies) which depends on the blood of their vertebrate hosts, including humans and most wild and domestic animals. While, thousands of flies like tachinids help in insect pest management which naturally causes destruction to pest insects from agro and forest ecosystems.

Indian agriculture is well recognized by varieties of crops grown but expected yield of the crops is not achieved so far because of the insect damage. Western Maharashtra is visualized as most advanced agricultural belt of India since, there are sufficient sources of water bodies and environment is very conducive for several agricultural crops. However, Sugarcane cultivation is most important feature of Western Maharashtra including Kolhapur, Sangli and Satara districts. The above three districts are famous for cultivation of additional crops such as pulses, cereals, oilseeds and vegetables. Vegetables and edible crops needs special attention for pest control since pesticidal use is not without danger to humans, other organisms and various ecosystems. Therefore, control strategies of pests without pesticides have prime importance and biological pest control can solve the problem of pesticidal side effects. The tachinids are biopesticides scattered in the nature as eco-friendly tool of pest management. They should be searched, conserved, protected and used in pest management.

Pest is harmful organism which causes at least 5 per cent damage to crop plants. But many pests severely damage to crops causing very higher losses sometimes 100 per cent (Sathe, 2015). Such a harmful insect pests from agricultural field can be eradicated by the use of different methods like cultural method, Biological methods, Chemical methods, Mechanical methods *etc*. Chemical control methods are widely used everywhere but, it causes pollution, health hazards and several other demerits like pest resistance, pest resurgence and secondary pest out breaks, *etc*.

Biological control method is keeping environment ecofriendly and is an alternative for chemical method. Biological control method is part of natural method and differs from other methods. Applied Biological control method means, "Manipulation of natural enemies by man to control the pests" (Van den Bosch *et al.*, 1982, Sathe, 2015). Natural enemies used in biological pest control are Predators and Parasitoids. Predators are animals that feed upon other animals (prey) and parasitoids are entomophagus insects that lives in or on and eventually kills a larger host.

Parasitoids are having different types like, idiobiont parasitoids immobilizes the host and prevent the development of it and koinobiont parasitoids allows the development of host while feeding upon it and may parasitize any host life stage. In turn, Koinobionts can be subdivided further into endoparasitoids, which grows inside body of the host, and ectoparasitoids, which develop outside the host body (Dawah, 2011).

Most parasitoids belong to the orders Hymenoptera or Diptera. Some dipteran flies specially tachinids(Family: Tachinidae) are good parasitoids as they kill the pests (host) and help for agricultural yield. Chatterjee and Misra (1974) reported 680 species of parasitoid of natural enemies of many insect pests. Insect predators and parasitoids annually provide US agriculture with an estimated $4.5 billion worth of pest control (Isaacs *et al.*, 2008) and this amount might be improved by altering the species composition and densities of non-crop plants in agricultural landscapes. Recently, Sathe (2014) described the different trends of biological pest control with the help of parasitoids and predators. He also described the role of tachinids in pest management. Many adult parasitoids and predators benefit from or require shelter, food and another hosts obtained from flowering plants (Root, 1973; Hickman and Wratten, 1996; Harmon *et al.*, 2000).Order Diptera includes several families of parasitoids, the largest family of them is Tachinidae. Tachinids are grouped into egg, larval and pupal parasitoids by their mode of attack.

The family Tachinidae is a member of the superfamily Oestroidea which is the second largest superfamily of Diptera, nearly 10,000 described species belonging to 1521 recognized genera occur in the world (O' Hara, 2008; Irwin et.al, 2003). Most tachinid flies are larger than a house fly and having more bristles with size from 2–20mm. In this family there is a huge diversity in shapes, colors and degree of bristling. Tachinids are second only to the parasitic Hymenoptera in diversity and represent the largest group of non-hymenopteran parasitoids and have ecological significance as insect parasitoids (Belshaw, 1994; Stireman, *et al.*, 2006).

Tachinid flies are beneficial to ecosystems found in several regions of the world. Though all are parasitoids, their reproductive strategies are different and far less understood than those of the parasitic Hymenoptera which usually lay eggs directly on or in the host body (Feener and Brown, 1997) but, some tachinids lay their eggs on host food plants and thus, they are different from Hymenopteran parasitoids. The biology of Tachinidae is fascinating and diverse especially in the adaptations of the female reproductive system, morphology of the immature stages and in the relationships between the larvae and their hosts (Ferrar, 1987).

The life cycle of Tachinid flies is very appealing that, they attack on an exposed host to undeveloped egg into the skin of host. Initially, a unembryonated egg is deposited directly on a host by the use of an extensible ovipositor. The egg shows thinner and flatter on under surface and thicker and convex on its upper surface which is attached to a host with a glue-like substance. A large oval planoconvex egg is having sticky substance on flat surface useful for adhesion on any part of host body. Developing embryo respire within by using minute pores *i.e.* Areopyle. When ready to hatch comes out by pushing off cap or operculum, eggs left to undergo inside host for embryonic development for several days and attend larval stage (O' Hara, 2008).

Tachinid flies show three larval instars. First instar search for their host and finally enters into host by their own self. Second instar feeds on non vital tissues. Third instar of larvae feed voraciously on host tissues and grows in size. Pupation observed in tachinid flies which takes place within tough shell (*i.e.* a puparium) of the third instar. The third instar crawls away from the host and pupae in the soil, but in some species pupation also takes place inside the host. The morphology of the spiracular discs shows variation in many species and provides valuable characters for the identification of tachinid species. An adult tachinid fly comes out of pupae (O' Hara, 2008).

Tachinids shows different colors from pale yellow to jet black, in some species shiny metallic blue or purple (*e.g., Gymnocheta*) or brightly metallic and different colored (*e.g.*, Rutiliini). Wings are clear but rarely darkened or sometimes bicolored in some species. Many species have four black longitudinal stripes on the thorax, which differs from the members of their closely related family Sarcophagidae that generally have three stripes. Chaetotaxy (arrangement of setae) on the head, thorax and abdomen is taxonomically very significant. Tachinid flies generally feeds upon nectar from flower and plant sap (O' Hara, 2008).

Tachinid flies are found in nearly all terrestrial environments and despite their abundance, diversity and ecological importance, relatively little is known about the evolutionary history, ecology and behavior of tachinids (Stireman *et al.,* 2006; O'Hara, 2010). Tachinidae is classified as Koinobiont because they do not kill their hosts immediately after parasitization (Askew and Shaw, 1986; Stireman *et al.,* 2006). Many Tachinid flies are parasitoids of some important pest insects as well as some millipedes, centipedes and spiders (Vincent, 1985; Wood, 1987; Williams *et al.,* 1990).Tachinid flies found in different habitat but life of them is endoparasitic which is complex one to find their perfect host. Generally, hosts of tachinid flies are plant feeding insects. Tachinid species can be mostly divided into two types, specialists and generalists. As the names mean, specialists attack one or a few host species and generalists attack multiple hosts (O'Hara, 2008).

Tachinid are having widest range of hosts, several of them belongs to order Hymenoptera, Coleoptera, Lepidoptera Orthoptera, Dermaptera and Embioptera. The majority of hosts of tachinids are Lepidoptera, mostly caterpillars which feeds on foliage. Lepidoptera represent the main hosts of the two tachinid subfamilies, Exoristinae and Tachininae and are also significantly attacked by the third sub-family Dexiinae. The smallest sub-family Phasiinae, attacks only Heteroptera and there is only one genus outside the Phasiinae (*Euthera* in the Dexiinae) that also attacks heteropteran bugs (O'Hara, 2008).

Tachinid flies parasitize larvae of Lepidoptera, larvae or adults of Coleoptera, larvae of Hymenoptera (sawflies and social Vespoidea) and Orthoptera. Dipteran parasitoids belong to the sub-family Exoristinae and Tachininae, which also parasitoids of the Orthoptera (crickets, grasshoppers), Blattaria (cockroaches), Mantodea (mantids), Phasmatodea (leaf and stick insects), Dermaptera (earwigs) and Embioptera (webspinners).Tachinid parasitize very rarely to other Diptera (Crosskey, 1977; Zeegers, 2007).

More than 100 species of tachinids have been used as biological control agents for agricultural and forest pests (Griener, 1988, O'Hara, 2008). Tachinid fly, *Lydella jalisco* successfully used as biocontrol agent against Mexican rice borer, *Eoreauma loftini* (Legaspi *et al.*, 1997). Order Orthoptera also serve as hosts for various tachinid genera. The tribe Acemyiini deposited egg directly on Grasshoppers. The genera of tachinine *Ormia* and *Glaurocara* are nocturnal, ovo-larvipositioning on long horned grasshoppers and crickets of families Gryllidae, Gryllotalpidae and Tettigonidae (Crosskey, 1967; Cade, 1975). Mantids are attacked by mainly Neotropical genus *Masiphya* (Guimares, 1966). Embioptera are hosts of many undescribed tachinids (Arnaud, 1963). However, Monteith (1955, 1956) showed that host plant was possibly the most important factor in attracting *Bessa harveyi* to its host. The tachinid *Ormia ochracea* is a parasitoid of eld crickets as internal parasitoid. After hatching the eggs and development larvae kills the host when they emerge and pupate into free-living adults (Adamo *et al.*, 1995 a; b).

Tachinid flies use both olfactory and visual senses to locate hosts and may be expected to be able to learn signal associated with both of these sensory modalities (Martin *et al.*, 1991; Monteith, 1955; 1956; Stireman, 2002). Tachinid flies always attracted towards host by its odour and movement which work as stimulants for tachinids. In different geographical regions, the tachinid fly uses different species as a host. It locates its hosts using the mating songs of male crickets and male parasitism rates can be as high as 80 per cent in some species (Cade, 1975). However, the relatively large eyes, the well-developed visual systems are characteristics of higher Diptera (Dethier, 1963; Strausfeld, 1970; Bushbeck and Strausfeld, 1997; Campbell and Strausfeld, 2001) and the demonstrated use of visual sense by tachinids suggest that learning of visual cues may be especially well developed in this group (Monteith, 1956; Weseloh, 1980; Tanaka and Honda, 1999a, b).

Tachinid biological control programs have been reviewed for many times, which include: the release of *Cyzenis albicans* from Europe against the winter moth, *Operophtera brumata* in Canada, release of *Lixophaga diatraeae* from Cuba and *Lydella minense* from Brazil against sugarcane borers, *Diatraea* sp. in the Caribbean, release of *Bessa remota* from Malaysia against the coconut moth, *Levuana iridescens* in Fiji and release of *Ceromasia sphenophori* from Papua New Guinea against the sugarcane weevil, *Rhabdoscelus obscures* (O'Hara, 2008).

Ormia depleta from Brazil has been implicated in the control of pest mole crickets, *Scapteriscus* sp. in Florida. Similarly, several species of *Trichopoda* from the southern United States and South America have been introduced recently into Hawaii, South Africa Italy and some other places for control of the southern green stink bug, *Nezara viridula* (O'Hara, 2008). However, little is known about diversity and taxonomy of Indian tachinid flies from different ecosystems especially forest and agriculture.

There is an evident loss of worldwide taxonomic expertise and relatively little is known taxonomically about several orders (Samways 2005). Similarly, very little study was carried on biodiversity and taxonomy of different orders of insects from India especially Western Maharashtra. Taxonomically, the *Tachinidae* is possibly the most complicated family of flies and perhaps because of this beneficial point the size of the family and their high position on the evolutionary tree of *Diptera* have received limited attention below the family level (Yeates *et al.*, 2007).

Keeping in view all above facts, the present taxonomic study was carried out with seasonal abundance, distribution, hosts and host plants of tachinids from both agro and forest ecosystems of western Maharashtra.

2

Review of Literature

At International scenario, several workers have been worked on Tachinid flies. Tothill (1924a; b) studied nearctic species of genus *Gonia* and genus *Fabriciella*. Aldrich (1925) described two new species of genus *Lixophaga*. Aldrich (1931) reported the notes on the genus *Cheatonodexodes* with one new species. Reinhard (1934) studied the New North American Tachinidae. Two winged flies of the genus *Doryphorophaga* from North America was reported by Reinhard (1935).

A revision on American species *Gonia* Meigen was done by Morrison (1940). The synopsis of the tachinid flies from genus *Tachinomyia* and description of new species was reported by Webber (1941). Brooks (1943) reviewed the Canadian species of *Ernestia*. Arnaud (1963) reported *Perumyia embiaphaga* new genus and species of Neotropical Tachnidae parasitic on Embioptera. Revision of the flies from the genus *Lespesia* in North America was carried by Beneway (1963). Verbeke (1963) studied the structure of the male genitalia in Tachinidae and their taxonomic value. Two new genera and species of tribe Eryciine from Australia was described by Crosskey (1967).

Crosskey (1967) revised oriental species of *Palexorista* Townsend of tribe Sturmiini. Kugler (1968) described six new species from Israel. Two new genera *Mesnilomyia* and *Palmonia* from Israel were described by Kugler (1972). New species of the genus *Thecocarcelia* T.T was described by Chao (1976). Crosskey (1976) studied taxonomic conspectus of the Tachinidae of the Oriental region. New species of the subtribe Peleteriina from China was described by Chao (1979). Chao and Shi (1980) described the genus *Blepharipa* Rond. from China. New species of *Hyalurgus* Brauer and Bergenstamm from China was described by Chao and Shi (1980).

Shima (1983) studied the new species of *Oxyphyllomyia* from Nepal with reference to the phylogenetic position of the genus. O'Hara (1983) described a new species, *Siphona kairiensis* from Australia. O'Hara (1983) described new species

Siphona kairiensis from Australia which is closely related to *S. gedeana*. Wood (1985) worked on taxonomic conspectus of the Blondeliini of North and Central America and the West Indies. New species of tachinidae from Israel and remarks on three little known species was reported by Herting (1986). Five new species of Exoristinae was described by Liu *et al.* (1986). New species of tachinid flies from Hengduan, mountains of China was described by Chao and Zhou (1987). Two new species of genus *Chetogena* Rond. from China was described by Chao and Chen (1987). Hiroshi Shima (1987) revised the Genus *Isosturmia* Townsend and described new species *I. pilosa.*

New genus and six new species of the tribe Goniini from China, Thailand and New Guinea were described by Shima and Chao (1988). Chao and Zhou (1989) described new genus *Chrysocosmius* Beggi of Tachinidae from China. New species of the genus *Parasetigena* from China was described by Mao and Chao (1990). Four new species of tribe Blondeliini from the Nanling Mountains of China was described by Yang and Chao (1990). Cantrell and Shima (1991) described new species of *Chaetophthalmus* Brauer and Bergenstamm from Papua New Guinea.

Shima *et al.* (1992) described the genus *Winthemia* from Yunnan Province, China. New species of the genus the *Zenilliana* from China was described by Sun and Chao (1992). New genus *Neophryxe* Townsend from China was described by Liang and Chao (1992). Richter (1992) described two new Palaearctic tachinid species of the tribe Elfiini. Tschorsnig (1993) described new species of the genus *Bithia* Robineau-Desvoidy from Tadjikistan and Kazakhstan. New species of the genus *Pexopsis* Brauer and Bergenstamm from China was described by Sun and Chao (1993). O'Hara (1993) revised species of the genus *Frontiniella* Townsend from America north of Mexico. Sun (1993) described two new species of the genus *Calozenillia* Townsend from China.

O'Hara (1994) revised tachinid species of the genus *Ceromya* Robineau-Desvoidy from Nearctic region. Richter and Zhumanov (1994) reported the genus *Goniophthalmus* for the first time from Middle Asia. New genus and species of the tribe Sturmiini from China was described by Sun and Chao (1994). Sun and Chao (1994) also studied on the Chinese species of *Phorocerosoma* Townsend. A new species *Lydella jalisco* was described from Mexico by Woodley (1994). Sun and Marshall (1995) described two new species of *Cylindromyia* Meigen with a review of the eastern Palaearctic species. Two new species of the genus *Istochaeta* Rondani from China was described by Liang and Chao (1995). A systematic study of the genus *Cavillatrix* Richter with ten new species was carried by Shima (1996). A new species *Borgmeiermyia paraguayana* of sub-family Exoristinae from Paraguay was described by Sehnal (1998).

Five new species of Tachinidae from Shanxi province, China was described by Liu *et al.* (1999). New and rare tachinid parasitoids of insect pests on the poplars (*Populus* spp.) in Bulgaria were reported by Georgiev (2000). Notes on two genera of *Microteryini* which are new to China descriptions were reported by Xu and Lou (2000). Nine Tachinidae species were recorded for the first time in Turkey by Kara (2001). A new species of the genus *Campylocheta* Rond. from Poland was described by Bystrowski (2001). Rayner and Raper (2001) identified and provided the keys for the species of tachinids and data on distribution and ecology was also added.

Review of the Chinese genus *Carcelia* Robineau-Desvoidy was given by Chao and Liang (2002). The species *Paracraspedothrix montivaga* Villeneuve and *Carcelia bombylans* Robineau-Desvoidy new to Britain were reported by Collins *et al.* (2002). The tachinidae and Oestridae of Madeira, with description of a new species were reported by Smit and Zeegers (2002). New species of the genus *Smidtia* Robineau-Desvoidy from China was described by Chao and Liang (2003).

A new species *Cestonia deserticola* was described by Richter (2004) from Turkmenia, which is similar to *C. canariensis* Vill. in having the frons wider than the eye. A new species of the tachinid genus *Thelaira* Rob.-Desv. from Tajikistan was described by Richter (2004). Zhang and Shima (2004) studied taxonomy of the genus *Dexia* Meigen from China. A new genus and species, *Rhinophoroides minutus*, of tribe Dexiinae from South Africa was described by Barraclough (2005). A new species of the genus *Phebellia* Robineau-Desvoidy from Finland was described by Bergström (2005).

Review of tachinid genera *Entomophaga* and *Proceromyia* was made by Tachi and Shima (2006). Liu *et al.* (2006) studied taxonomy of the tribe Blondeliini from China. Raper (2006) reported *Carcelia laxifrons* Villeneuve for the first time from Britain and a revised the key to the British *Carcelia* species. Bystrowski (2006) reported tachinid species *Ramonda latifrons* Zetter for the first time to Polish fauna.

Cerretti and Wyatt (2006) described new species of the genus *Eomedina* Mesnil from Namibia. Revision and systematic placement of the genus *Prospalaea* Aldrich was done by Nihei (2006). Nihei and Pansonato (2006) revised of *Prophorostoma* Townsend with the description of a new species. A systematic study of the genus *Dinera* Robineau-Desvoidy from Palaearctic and Oriental regions was done by Zhang and Shima (2006). Jones (2006) recorded *Catharosia pygmaea* Fall. for the first time from Kent. Zhang *et al.* (2006) described new species and a new genus *Leptothelaira* from China.

Bergstrom (2007) described tachinid species, *Loewia erectan* from Fennoscandia and Poland. Bergstrom (2007) described new species of *Nilea* Robineau-Desvoidy with notes on the genus and a key to the North European species. Zhang *et al.* (2007) recorded the genus *Atylomyiain* for the first time from China with two new species. Taxonomy of the genus *Phebellia* R.D from China was described by Chao and Chen (2007).The tachinid fly *Phasia hemiptera* Fab. was first time reported from Norway by Gammelmo and Sagvolden (2007). A new species of tribe Blondeliini, *Meigenia fuscisquama* was described from Jilin and Liaoning region of China by Liu and Zhang (2007). Cerretti and Barracloug (2007) described new genus and species, *Anomalostomyia namibica* of Afrotropical Tachinids. Cerretti and Tschorsnig (2007) described two new species of *Siphona* Meigen from Sardinia and Morocco. A new species of the genus *Plesina* Meigen from the Mediterranean was described by Cerretti and Tschorsnig (2008).

Taxonomy of subgenus *Tachina* Meigen from North China was studied by Hao *et al.* (2008). Similarly, taxonomy of the genus *Frontina* Meigen from Korea was studied by Lee and Han (2008). New species of the genus *Paravibrissina* from Southeast Asia and South Pacific was described by Shima and Tachi (2008). A new species of *Leschenaultia barbarae* was described from Venezuela by Toma (2008), which was closely runs towards the species, *Leschenaultia bicolor* Macq. Two tachinid species, *Leskia erevanica* Rich. and *Tachina grossa* Linn. are recorded for the first time from Turkey by Kara and Aksu (2008). A new genus and species, *Lobomyia neotropica* was described from Colombia, Brazil, Costa Rica, Mexico and Trinidad by Woodley *et al.* (2008). Lim and Han (2008) redescribed two closely resembling of the genus *Linnaemya* species from Korea. Perry (2008) recorded *Opesia grandis* Egger for the first time from Cambridgeshire.

Jones (2009) recorded *Gymnosoma rotundatum* Linn. for the first time from urban London. Falk *et al.* (2009) recorded *Leucostoma anthracinum* Meigen for first time to Britain. Gheibi and Ostovan (2009) studied fauna of tachinid flies in Fars Province, Iran. Shima and Tachi (2009) described a new species of the genus *Setalunula* Chao and Yang from Japan. Two species of the genus *Phasia* Latreille was newly recorded from Korea by Cha and Han (2009). A taxonomic revision of the genus *Metadrinomyia* with descriptions of two new species has been made by Byun and Han (2009).

A data on distribution of 40 species belonging to the subfamilies Exoristinae and Tachininae were collected from Fars province, Iran was recorded by Gheibi *et al.* (2009) and systematic study of the genus *Rossimyiops* Mesnil by Cerretti *et al.* (2009). Gheibi *et al.* (2009) reported of six tachinid flies for the first time from Iran. Shima *et al.* (2010) made a description of a new genus and six new species of Tachinidae from Asia and New Guinea. A taxonomic note on the genus *Borgmeiermyia* Townsend with the first host record has been made by Nihei and Toma (2010). A new species *Besseria prophetarum* was described from Israel by Cerretti *et al.* (2010).

Taxonomy of the subgenus *Servillia* Robineau- Desvoidy from North China was studied by Chi *et al.* (2010). Draber and Nowakowski (2010) reported *Euthera fascipennis* Loe. for first time from Tunisia, with a description of female and the puparium. Ziegler (2010) revised the genus *Germaria* Robineau-Desvoidy from Greece, with descriptions of two new species. Ten species of Tachinids were recorded from Saudi Arabia by Dawah (2011), out of which eight species were reported for the first time. Taxonomic redescription and biological notes on *Diaugia angusta* Pert. was provided by Nihei and Pavarini (2011).

O'Hara (2011) prepared checklist of world genera of the tachinidae and their regional occurrence. Karagoz, *et al.* (2011) recorded tachinid species, *Microphthalma europaea* Egger for the first time from Turkey. Cerretti *et al.* (2012) described a new genus *Neoethilla* of the Ethillini from the New World. Nunez and Couri (2012) redescribed *Uruleskia* Townsend of the type-species. O'Hara (2012) reviewed the genus *Euthera* from North America with the description of a new species. Wang and Zhang (2012) described new species of the genus *Neaera* Robineau-Desvoidy from Sichuan, China. Zhang and Fu (2012) described three new species of *Dinera* Robineau-Desvoidy from China. While, Zhao *et al.* (2012) described one new species of the genus *Kuwanimyia* Townsend from Zhanjiang, China. Three new species of *Estheria* Robineau-Desvoidy from the Mediterranean were also described by Cerretti and Tschorsnig (2012).

A New genus and a new species of Tachinids *Aesia acerbiana* belongs to the tribe Blondeliini from Wrangel Island was described by Richter (2012). Gilasian *et al.* (2013) reviewed the genus *Cylindromyia* Meigen in Iran, with description of two new species and the newly discovered male of *C. persica* Tschorsnig. A Newly recorded genus and species of Tachinids, *Sericozenillia albipila* Mesnil from China was described by Qi *et al.* (2013). Zeegers and Jahromi (2015) recorded genus *Blepharella* from the western Palaearctic region and the species *Blepharella setigena* Cort. for the first time from Iran.

Dominant workers on taxonomy of tachinids from India refer to Das (1993), O' Hara (2008), Lahiri (2003, 2006) *etc.*

Das (1993) described new species of *Turanogonia* Rogdendorf from Darjeeling district West Bengal. Lahiri (2003) studied Tachinidae fauna of Sikkim state. Lahiri (2006) studied Tachinidae fauna of Nagaland state. Sathe (2012) studied the biodiversity of tachinids from western Maharashtra. Sathe *et al.* (2014) also studied diversity of tachinids from agroecosystems of Kolhapur districts. The review of literature indicates that very little information is known about taxonomy of tachinid flies from Western Maharashtra. Hence, present topic was selected.

3

Collection and Preservation of Tachinids

Materials and methods is very important component in research. Material facilitates for technical support and methods help for getting superior results. Any minor change in material or methodology, results in extreme change in the results, may be positive or negative. Therefore, following materials and methods were adopted for study of taxonomy and seasonal abundance of Tachinid fly.

1) Insect Collecting Net (Figure 4)

Insect collecting net made up of iron handle, 70 cm long with metal ring of 22cm diameter; with nylon net of length 60cm forming bag was used for collecting flies.

2) Plastic Containers (Figure 5)

Plastic containers having capacity of 2 liters were used for handling and carrying the caterpillars as hosts and flies. These containers were perforated with tiny holes for ventilation for collected insects.

3) Specimen Bottles (Figure 6)

Specimen bottles of size 2.5× 6cm were used for collection and carrying the tachinid flies.

4) Camel Hair Brushes (Figure 7)

Camel hair brush of No. 4 and 8 have been used for regular cleaning of preserved tachinid specimens.

5) Insect Spreading Board (Figure 8)

Insect Spreading boards were used for proper setting and pinning of tachinid flies for taxonomic studies.

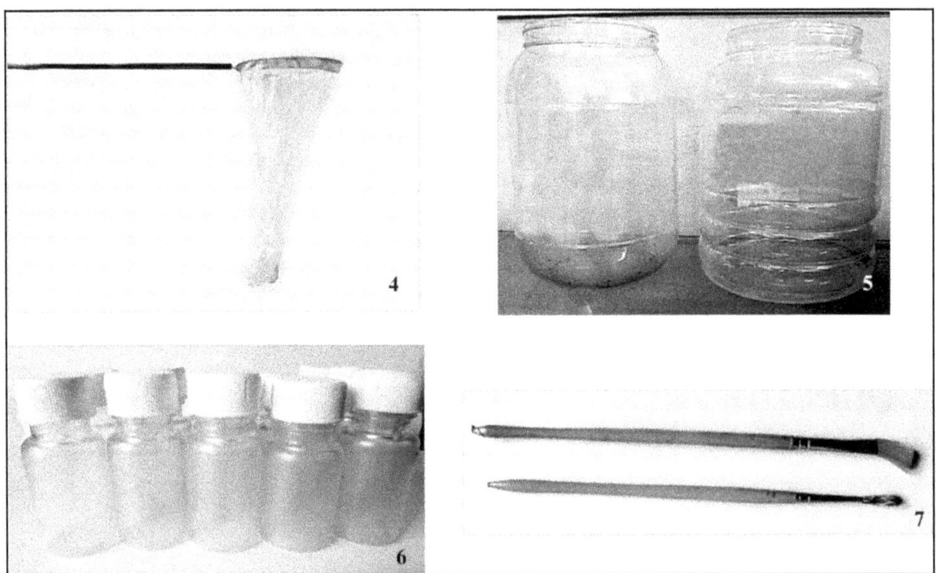

Plate 1: Figure 4: Insect Collection Net; Figure 5: Plastic Containers; Figure 6: Specimen Bottles; Figure 7: Camel Hair Brushes.

6) Oven (Figure 9)

Oven size 4× 3 feet (height and width) has been used for drying the pinned specimen of tachinid flies at 35°–40°C temperature.

7) Insect Storage Box (Figure 10)

Clean insect storage box having size of 30 × 45cm was used for proper preservation of dried tachinid flies.

8) Hand Lens (Figure 11)

Hand lens with 10cm diameter was used for study of morphological characters of tachinid species.

9) Glass Cage (Figure 12)

Glass cage of size 30x30x30 cm (length, height and width) having wooden case and glass walls on three sides. One side door was covered by muslin cloth handling insects.

10) Compound Microscope (Figure 13)

Morphological and taxanomical studies of tachinid fly were carried by using compound microscope with 10x, 45x objects and 10x eyepiece.

11) Camera (Figure 14)

Insect photography was made in the field and laboratory with the help of Canon camera 600D (made in Japan). For close up photographs of tachind flies +1, +2, +4 micro lens and 18 x 55 mm macro lens was also used.

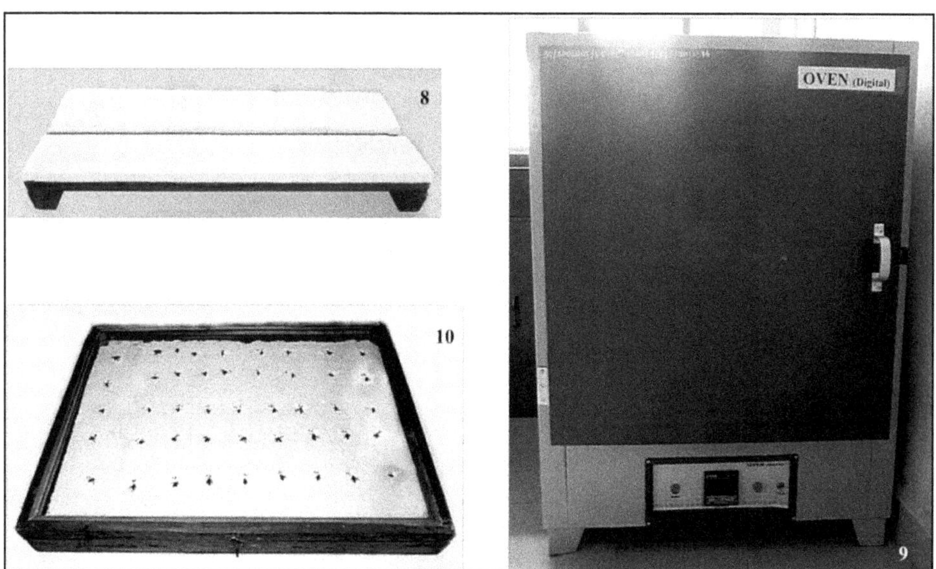

Plate 2: Figure 8: Spreading Board; Figure 9: Oven; Figure 10: Insect Storage Box.

12) Slides and Cover Slips (Figure 15)

Slides and cover slips were used for preparing slides of body parts of tachinid flies for taxonomic and morphological studies.

13) Entomological Pins (Figure 16)

Entomological pins of No. 00 have been used for pinning the tachinid flies.

14) Card Sheet Paper

White colored card sheet papers were used for carding and numbering the tachinid flies.

15) Chemicals

Following chemicals were used for the study of tachinid flies.

a. Chloroform for killing the collected tachinid specimens.
b. Naphthalene balls were used for avoiding fungal infestation to specimen.
c. 70 per cent alcohol was used for wet preservation.
d. Alcohol grades 30 per cent, 70 per cent, 90 per cent, absolute.
e. Xylene
f. DPX

Methods

1. Collection and Preservation of Tachinid Flies

Tachinid flies have been collected from study area of Kolhapur, Sangli, Sidhudurg and Satara districts with the help of insect net and specimen bottles.

Plate 3: Figure 11: Hand Lens; Figure 12: Glass Cage; Figure 13: Compound Microscope; Figure 14: Camera (Canon); Figure 15: Leg Slides; Figure 16: Entomological Pins.

The collected tachinid flies were anesthetized by using chloroform and preserved in the box by appropriate pinning. The pinned tachinid flies were dried in the oven at 55°- 60°C for 1 week and only intact and scientifically pinned insects were preserved in insect boxes. Naphthalene balls were used for avoiding infections to preserved insects in boxes. Tachinids were pinned from the ventral side from mesothorax. Pins were pointed at both ends. The pin was not allowed to emerge from dorsal surface for avoiding damage to chaetotaxy.

2. Taxonomy of Tachinid Flies

Taxonomical studies of tachinid flies were carried with the help of compound microscope/hand lens for further description. Measurements of tachinid body parts were taken in mm in the description. Colored photographs of tachinid flies have also been taken for supportive evidence for the characters. The taxonomical descriptions of the tachinid flies have been made as per the terminology and keys of Crosskey (1976), Rayner and Raper (2005). The described species are being preserved in the department of zoology, Shivaji University, Kolhapur and will be deposits in appropriate institute if needed.

Genitalia

Genitalia of both males and females were directly photographed with the help of Canon 600 D.

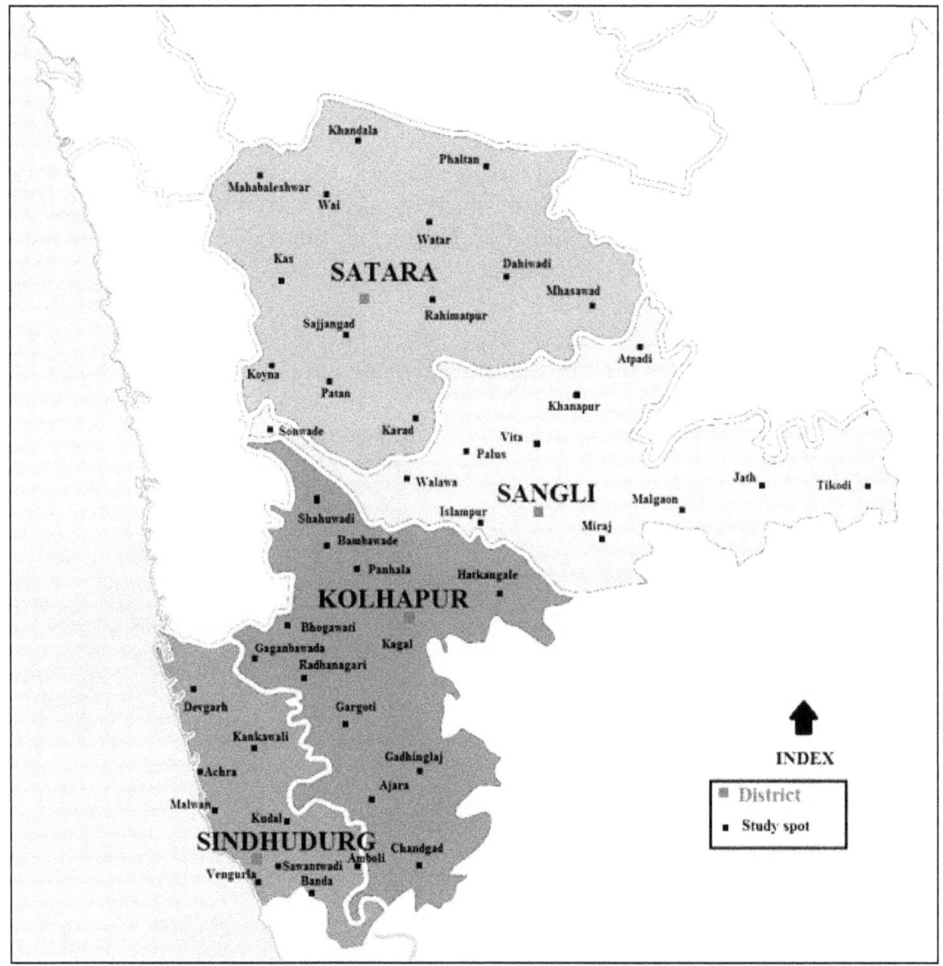

Figure 17: Map of Districts: Kolhapur, Sangli, Satara and Sindhudurg Showing Collection Spots.

3. Host Records

Host records of tachinids have been prepared by rearing of lepidopterous larvae in laboratory. Continuous observations were made on life stages of tachinids for adult emergence. The newly emerged adults were then identified. Both, hosts and tachinids were identified by consulting appropriate literature cited in bibliography.

4. Distribution of Tachinid Flies

Distributional records of adult tachinid flies have been made from study area by spot observations of the species at 15 days interval at morning and evening.

5. Seasonal Abundance

Seasonal abundance of tachinid flies have been studied by spot observations and collections of the species from different study spots at 15 days intervals throughout the year, during entire tenure of study by adopting one man one hour search method.

Study Area

The study spots which includes four districts of Western Maharashtra namely Kolhapur, Sangli, Satara and Sindhudurg are shown in Figures 1–3 and 17–25.

Plate 4: Study Spots. Figure 18: Rice Field; Figure 19: Sugarcane Field; Figure 20: Carrot Field; Figure 21: Sunflower Field.

Plate 5: Study Spots. Figure 22: Shahuwadi; Figure 23: Hatkanangale (Ramling); Figure 24: Amboli; Figure 25: Panhala.

Kolhapur district is extended in between 15° to 17° (N) latitude and 73° to 74° (E) longitude and surrounded by Belgaum (Karnataka) at South and East, Ratnagiri and Sindhudurg districts at West and Sangli district from North. The area of Kolhapur District is about 7,685 km^2. Out of which, 417 sq. km. is protected forest area and 563 sq. km. is reserved forest area which covers a total of 22 per cent forest area and contains Subtropical evergreen, moist deciduous and semi-evergreen and dry deciduous forests.

The subtropical evergreen forest shows most important trees such as Yellow Myrobalan, Java plum, Jack fruit, ficus tree, etc. Moist deciduous forest and dry deciduous forest typically shows reserve and protected forest known for firewood and grass as a main marketable products from this forest. The majority area of western ranges is under thick forest with red soil. In Western Maharashtra Panchganga, Bhogawati, Dudhganga, Ghatprabha, Vedganga, Warana and Hiranyakeshi are main rivers which flow towards east. The Kolhapur district receives its major (average 1600mm) rainfall from the South west monsoon from June to October. Kolhapur district covers 12 tahasils. Out of which Chandgad, Ajara, Gadhingalaj, Radhanagari, Shahuwadi, Gaganbawada comes under hilly region of Western Ghats which are important from the point of view to study tachinid diversity from forest and agroecosystem.

Sangli district lies in between 16° 45′ to 17° 38′ (N) latitude and 73° 42′ to 75° 40′ (E) longitude which is bounded by Bijapur (Karnataka) at East and Kolhapur district on West. Satara and Solapur district partially surrounded North region. The total area of this district is 8,578 km^2. The climatic condition is hotter and drier towards the east and humid towards the west, temperature rises between 31.2°C to 41.5°C in May-July and falls 10.3°C to 21.5°C in November- January months. Study spots refer to Sangli district are Battis shirala, Miraj, Atapadi, Kundal, Tasgaon, Islampur, Vita, Sagareshwar and Chandoli.

Satara district lies in between 17° 5′ to 18° 1′ (N) latitude and 73° 33′ to 74° 74′ (E) longitude with a total area of 10,484 km^2 and bounded by Solapur at East, Ratnagiri district at West and Pune at North side and subdivided into tahsils namely Medha, Satara city, Wai and Mahabaleshwar which comes under hilly region from Western Ghats. Ajinkyatara, Mahabaleshwar, Panchgani, Yavteshwar, Salpani hills, Vankusawade and Bhairavagad hills comes under Satara district having altitude rage of 900 and 2150m from sea level. Koyana wild life sanctuary and Koyana dam also comes under Western Ghats. In plain region of Satara district important crops like paddy, sugarcane, wheat, maize, soya bean, etc are cultivated.

Sindhudurg district lies in between 15° to 17° (N) latitude and 73° to 75° (E) longitude. This district is surrounded by Kolhapur at East, Ratnagiri district at North. Goa and Karnataka partially surrounds south side. The area of this district is 5,207 km^2. It shows maximum area of Western Ghats. Sindhudurg district have Dodamarg, Sawantwadi, Vengurla, Kudal, Malvan, Kankavli, Devgad and Vaibhavwadi tahsils. The chief crops produced in Sindhudurg are rice, mango, coconut, nachani, cashew nut, jackfrui, groundnut and variety of spices. Amboli, Kunkeshwar, Mahadevgad hills comes under Sindhudurg district showing altitude of 690 m from sea level. Heavy rain fall (about 3240 mm) occurs in many places of this district. West side of Sindhudurg district is surrounded by sea shore and covers large area. In plain region of Sindhudurg Paddy, wheat, ground nut, soya bean and certain fruit plants like coconut, guava, papaya, *etc.* are cultivated.

On the basis of diversity, geographical and environmental conditions Satara, Sangli, Kolhapur and Sindhudurg districts were selected from Maharashtra as study areas. Tachinids have been studied from both agro and forest ecosystems of above said districts.

4

Taxonomy of Tachinids

The word taxonomy is derived from the Greek words, taxis (=arrangement) and nomos (=law), deals with the study of collection, identification, nomenclature, description and classification of organisms" (Mayer, 1942). The value of taxonomy in classification system is that it allows biologists to distinguish relationship between different organisms.

Study of diversity of plants and animals is very important for conservation and development of living organisms. Diversity is studied with respect to physiology, anatomy, genetics, taxonomy, biotechnology and barcoding etc. Therefore, in the present work was objected to study taxonomy of tachinids.

The exact relationship of systematics and classification to taxonomy varies from source to source. Davis and Heywood (1963) describe taxonomy as a system of arranging and interpreting information. According to Narendran (2000) taxonomy is usually the basis for a number of significant studies and is considered an important basic science giving skeletal support to many other sciences.

According to Judd *et al.* (2007) taxonomy is the theory and practice of grouping individuals into species, arranging species into larger groups and giving names to those groups, thus producing a classification. Simpson (2010) Says that, taxonomy is a field of science that includes description, identification, nomenclature and classification.

Taxonomy traditionally covers three areas *i.e.* Alpha taxonomy, beta taxonomy and gamma taxonomy. Alpha taxonomy is the stage at which the species are identified and described, while beta taxonomy refers to the arrangements of the species into a natural system of lower and higher level on the basis of relationship and gamma taxonomy is the analysis of intra specific variations, ecotypes, phylogeny *etc.* (Disney, H., 2000).

Taxonomic study is essential in many other fields *viz.*, agriculture, evolutionary biology, fisheries, medicine, public health, wild life management, etc. It is simple tool for understanding the biodiversity and simultaneously, helps for conservation of many species. The study of taxonomy further helps for better understanding of their origin, diversification and the evolutionary history by combining morphological, molecular and developmental information.

Tachinid flies are good biocontrol agents available in different ecosystems. Hence, they have been studied with respect to taxonomy, hosts and distribution.

In past, Aldrich (1925,1931), Crosskey (1976, 1984), O'Hara (1983, 1984, 2002, 2010, 2012), Wood (1985), Hiroshi Shima (1987, 1996), Woodley *et al.* (1994, 2008), Sehnal (1998), Barraclough (2005), Toma (2008), Byun and Han (2009), Gheibi *et al.* (2009), Richter (2008, 2009, 2012), Tunca *et al.* (2009), Cerretti (2009), Novotná *et al.* (2009), Bentley and Raper (2010), Cantrell and Burwell (2010), Dawah (2011), Karagöz *et al.* (2011), Cerretti *et al.* (2012), Wang and Zhang (2012), Gilasian *et al.* (2013), Tachi (2013), Wang *et al.* (2014, 2015), Desai et. al (2015) etc. have contributed on tachinid fly taxonomy from different parts of the world.

Adult tachinid flies have been collected from study spots of Sangli, Satara, Kolhapur and Sidhudurg district (Figure 17) with the help of insect collecting net at morning and evening. The collected specimens were pinned, dried at 55-60° C and kept in insect storage box with naphthalene balls. The taxonomical studies were carried by consulting literature of Crosskey (1976), Wood (1987) and Rayner and Raper (2005). The details of methodology are given in the chapter 3, materials and methods.

Morphological Considerations

The body of tachinid fly is divided into head, thorax and abdomen (Figures 26–36).

Head (Figure 28)

Head is well differentiated and separated from the thorax by a short membranous region. Head is variable in proportion generally higher than long with slanting frons, long antennae and face is larger than back of head. Short antennae and small face in dichoptic flies. Two narrow strips between the ptilinal and frontogenal sutures are called *facial ridges*. It bears more or less prominent, thick and stout bristle, named *vibrissa*. It is having other setae known as supravibrissal bristles. Frons is the entire part between the eyes, the gena, the vertex, and the frontoclypeal suture. Ocelli are arranged on a plate, called *ocellar triangle* or *ocellar plate*. **Eyes** present on lateral region of head with more or less hairs in between facets, eyes may be rounded, oval mostly dichoptic in nature. **Gena** is paired sclerite that occupies the area of head below the eye. **Antenna (Figure 29)** consists of three segments *i.e.* scape, pedicel and flagellomere (post pedicel). Scape is very smaller than rest of all segments. Pedicel is more or less globular and elongated in shape. Post pedicel bears filiform appendages on external part called as arista.

Thorax (Figures 30 and 31)

The thorax is the second morphological region of the body. The exoskeleton of each thoracic segment shows junction of four plates: the dorsal tergum and the ventral sternum and two lateral pleura. Tergum is divisible into prescutum, scutum and scutellum. The pleural region is characterized by the presence of the spiracles. At lower region on pleuron thoracic segments are present *i.e.* proepisternum, mesoepisternum, metaepisternum. Horizontal suture divided episternum into two sclerites *i.e.* anepisternum and katepisternum. Another suture present behind horizontal suture also divide epimeron into two sclerites *i.e.* anepimeron and katepimeron. Sternum is ventral plate that covers the exoskeleton of each thoracic segment. **Pronotum** dorsally divided into two sclerites *i.e.* antepronotum and postpronotum. Out of which, postpronotum are well visible into two lateral expansions called humeral callus. The lateral region (propleuron) divided into two sclerites, the proepisternum (anterior) and the proepimeron (posterior).

Mesonotum dorsally shows four parts, prescutum, scutum and scutellum. Prescutum is short and shows post humerals, presutural, Acrostichal, Dorsocentral, Intralar, Supralar, Postalar setae. Scutum dorsally shows anterior and posterior swellings *i.e.* prealar callus and postalar callus. Post alar ridges present ventrally to postalar callus. Scutellum placed immediately behind scutum. Scutellar bristles (*i.e.* Basal, lateral, subapical and apical) are present at posterior border of scutellum.

Postnotum located behind the scutellum also called as postscutellum. It consists of three sclerites *i.e.* one mediotergite (subscutellum) and paired laterotergites. Laterotergites are divided into small plates *i.e.* anatergite and katatergite. Mesopleuron lies at middle region of pleuron. A pleural suture goes through the mesopleuron from the base of the coxa to the insertion of wing and divides mesopleuron into two pleural plates, called episternum (anterior) and epimeron (posterior). Episternum is divided by anapleural suture into dorsal anepisternum (mesopleuron) placed between the anterior spiracle and the insertion of wing and ventral katepisternum.

Wings (Figure 32)

Fore wing and hind wings are present but, hind wings got sharply reduced called as, Halter. Fore wings are divisible into three areas, axillary area which is the proximal end, basal stalk a transitional area and blade a large area. Axillary area shows sclerites like, tegula or costal sclerite, basicosta or humeral sclerite, subcostal sclerite, first axillary sclerite, second axillary sclerite, third axillary sclerite. Basal stalk shows the proximal and distal median sclerites. The stalk encloses the bases of longitudinal veins *i.e.* costa (c), subcosta (sc), radius (r), posterior media (m), cubitus (cu), and anals (a) and the humeral (h) cross vein. The major membranous areas of the base are covered by the basal costal cell (bc), the alula, and the calypteres (Lower and upper).

Legs (Figure 33)

The legs of tachinids are appendages consisting of three pairs, the forelegs are called prothoracic legs, the mid legs are mesothoracic legs and the hind legs are

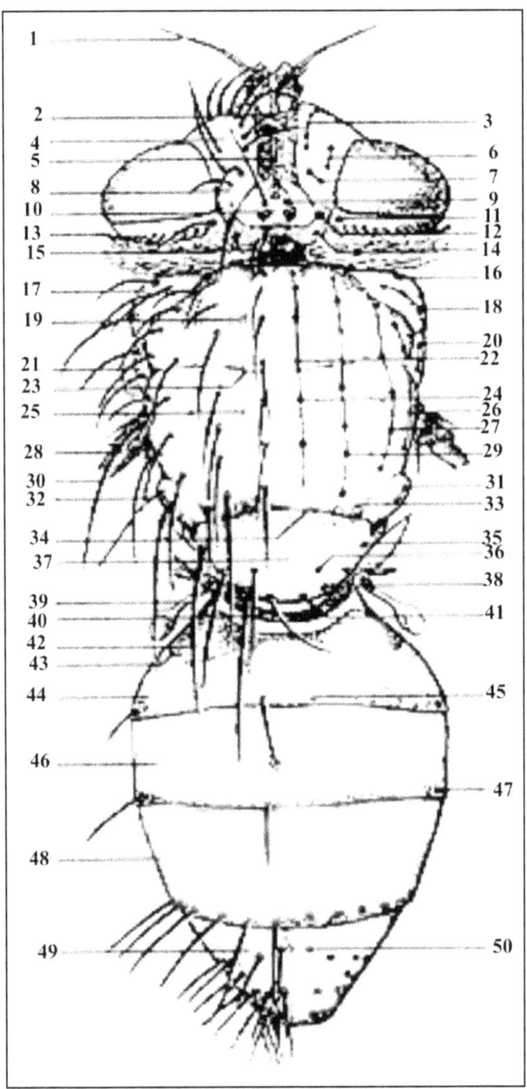

Figure 26: Morphological Features of Tachinid Fly (Dorsal View).

1. Arista; 2. Lunula; 3. Frontal Bristle; 4.parafrontalia; 5.Frontal vita; 6. Outer orbital bristles; 7. Inner orbital bristles; 8. Vertex; 9. Ocellar bristles; 10. Ocellar triangle; 11. Outer vertical bristle; 12. Inner vertical bristle; 13. Back of Head; 14. Occiput; 15. Post Ocellar bristles; 16. Humeral bristles; 17. Humeral callus; 18. Basal humerals; 19. Prescutum; 20. Notopleural bristles; 21. Presutural bristles; 22. Acrostichal bristles; 23. Suture; 24. Post sutural acrostichal bristles; 25. Scutum; 26. Supra alar bristles; 27. Intraalar bristles; 28.Wing base; 29. Dorso central bristles; 30. Prealar bristles; 31. Postalar callus; 32. Postalar bristles; 33. Post scutum Bristles; 34. Interscutelar suture; 35. Basal bristles; 36. Discal scutelar bristles; 37. Scutellum; 38. Subapical scutellar bristles; 39. Halter; 40. Excavation; 41. Apical scutellar bristles; 42. Lateral scutellar bristles; 43.Tergite-1; 44.Tergite- 2; 45. Dorsomarginal bristles; 46.Tergite -3; 47. Lateral marginal bristles; 48.Tergite- 4; 49.Tergite -5; 50. Discal median bristles.

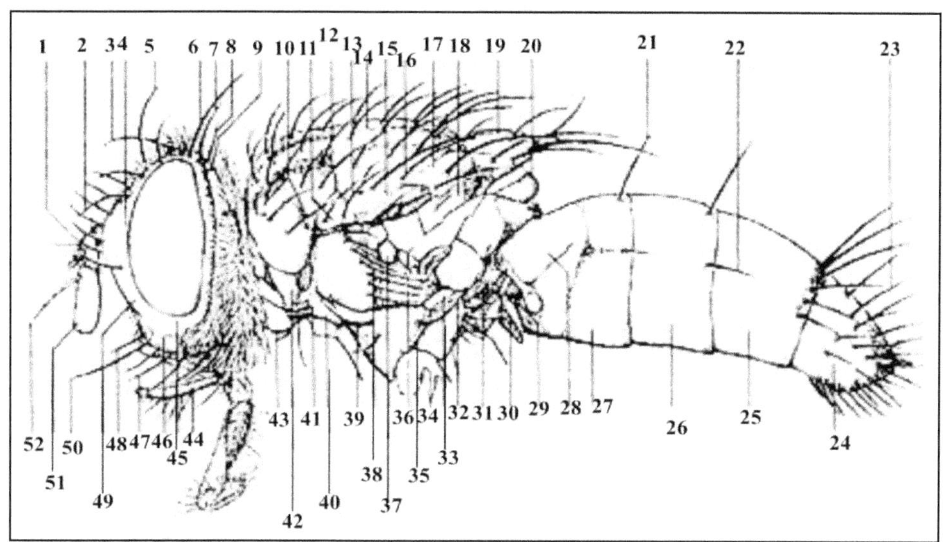

Figure 27: Morphological Features of Tachinid Fly (Lateral view).

1. Pedicel; 2. Proclinate orbital bristle; 3. Ocellar bristle; 4. Parafrontalia; 5. Reclianate orbital bristle; 6. Inner vertical bristle; 7. Outer vertical bristle; 8. Posto ocellar bristle; 9. Post ocular bristle; 10. Humaral callus; 11. Notopleuron; 12. Suture; 13. Notopleural bristle. 14. Scutum; 15. Supraalar bristle; 16. Prealar bristle; 17. Post alar callus; 18. Supra squamal ridge; 19. Scutellum; 20. Postscutellum; 21. Marginal bristle; 22. Lateromarginal bristle; 23. Median Discal bristle; 24. Tergite-5; 25. Tergite-4; 26. Tergite-3; 27. Tergite-2; 28. Tergite-1; 29. Halter; 30. Posterio ventral declivity of thorax; 31. Spiracle; 32. Hypopleural bristle; 33. Barette; 34. Coxa; 35. Hypopleuron; 36. Pteropleuron; 37. Bulbus; 38. Mesopleural bristle; 39. Sternopleural bristle; 40. Proepimeron; 41. Prostigmatic bristle; 42. Propleuron; 43. Propleural setae; 44. Palp; 45. Gena; 46. Genal dilation; 47. Peristomal setae; 48. Perivibrissal bristle; 49. Parafacials; 50. Vibrissa; 51. First flagellomere; 52. Arista.

metathoracic legs. Each leg consists of segments like coxa, trochanter, femur, tibia, tarsus and pretarsus. Femora each leg shows preapical anterio dorsal setae, preapical anterio ventral setae, preapical posterio dorsal setae, preapical posterio ventral setae.

Abdomen (Figure 34)

Abdomen shows five complete tergites in lateral view. Generally, in males segment 7+8 fused with tergite 6. Tergite 1 and 2 are fused to form syntergite 1+2. It also shows excavation as mid dorsal depression. Abdominal tergites usually show marginal bristles, discal bristles, laterodiscal bristles, *etc.* Abdominal sternites are also five in number but, many times it is overlapped by ventral edges of tergites.

Genitalia

Male Genitalia (Figure 35)

Retracted within 5th abdominal tergite. Syntergosternite 7+8 at posterior shows epandrium. Ventrally it shows attachment of Pregonite, Postgonite, Phallus, Basophallus and distiphallus, Surstylus and Cercus.

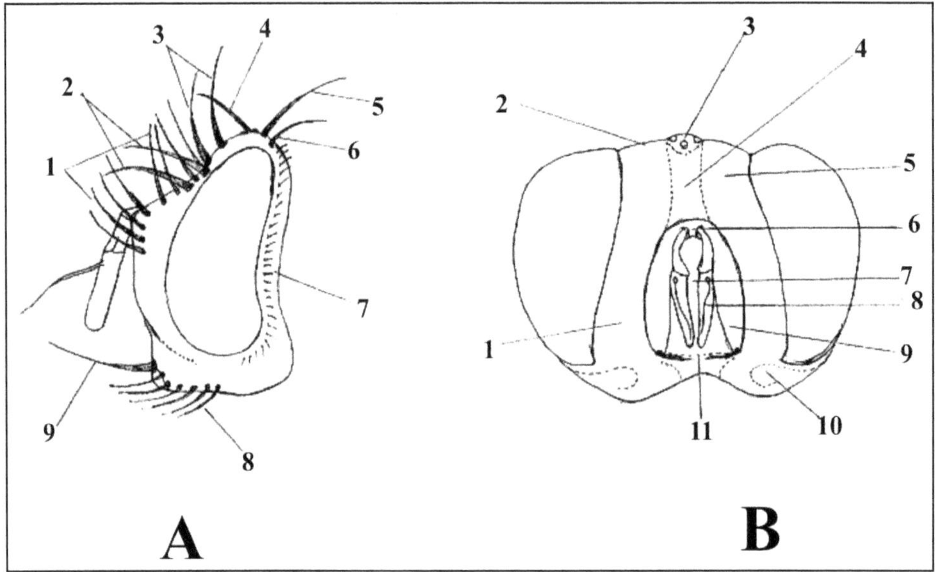

Figure 28: A: Morphological Feature Head of Tachinid Fly (Lateral view).

1. Frontal bristles; 2. Proclinate orbital bristle; 3. Reclinate orbital bristle; 4. Ocellar bristle; 5. Inner vertical bristle; 6.Outer vertical bristle; 7. Postocular bristle; 8. Peristomal bristle; 9; Vibrissa.

B. Morphological Feature Head of Tachinid Fly (Frontal view)

1. Parafacial; 2. Vertex; 3. Ocellar triangle; 4. Frontal vitta; 5. Parafrontal; 6. Lunula; 7. Face; 8. Antenna; 9. Facial ridge; 10. Genal dilation; 11. Epistome.

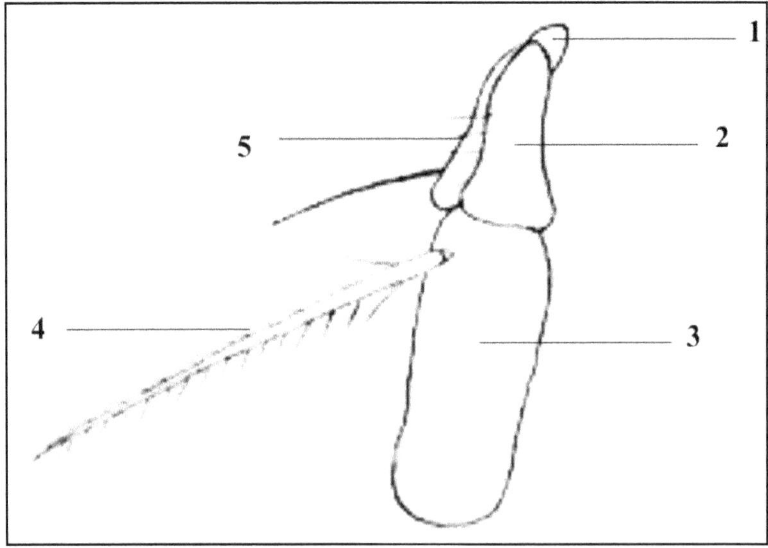

Figure 29. Morphological Feature of Antenna.

1. Scape; 2. Pedicel; 3. Flegellomere; 4. Arista; 5. Antennal seam.

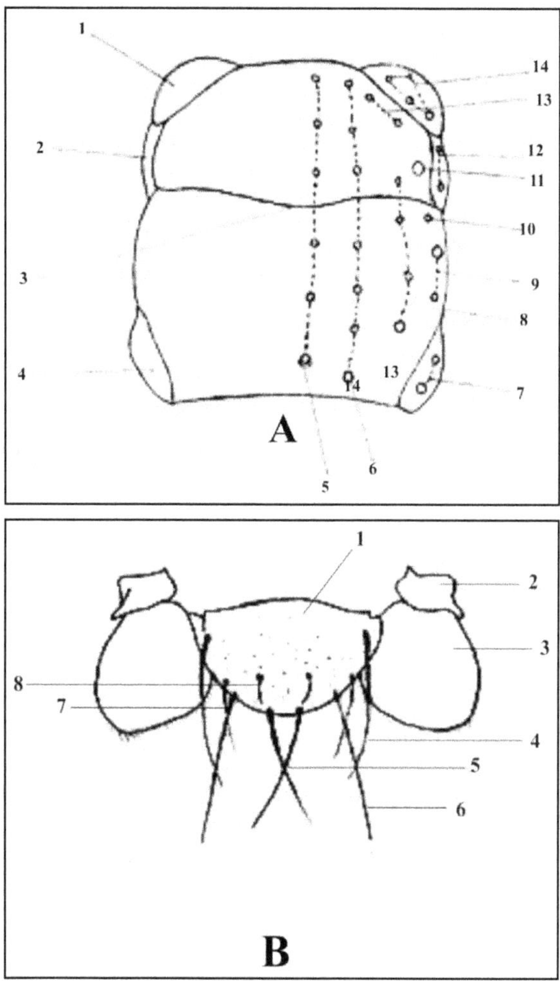

Figure 30: Morphological Features.

A. Thorax: 1. Humeral callus; 2. Notopleuron; 3. Transverse suture; 4. Post alar callus; 5. Acrostichal bristle; 6. Dorsocentral bristle; 7. Post alars; 8. Intra alars; 9. Supra alars; 10. Prealars; 11. Presutural bristle; 12. Notopleurals; 13. Posthumerals; 14. Humerals.

B. Scutellum: 1. Scutellum; 2. Upper calypter; 3. Lower calypter; 4. Basal scutellar bristle; 5. Apical scutellar bristle; 6. Subapical scutellar bristle; 7. Lateral scutellar bristle; 8. Discal scutellar bristle.

Female Genitalia (Figure 36)

The female terminalia (also called hypopygium or postabdomen) lies from sixth or seventh posterior segments. Ovipositor is slender tube composed of the posterior segments retracted into segment 5th or 6th segment. Cerci are small pad like structure.

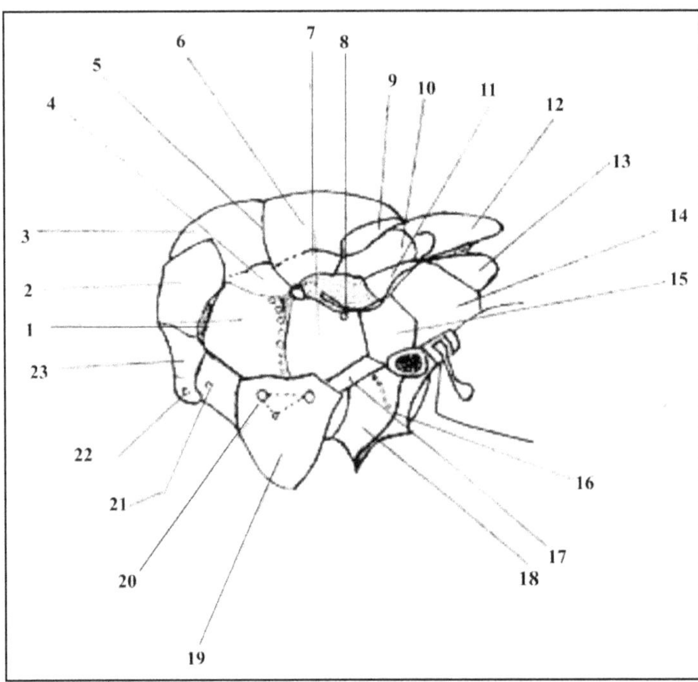

Figure 31: Morphological Features of Thorax (Lateral View).

1. Mesopleuron; 2. Humeral callus; 3. Prescutum; 4. Notopleuron; 5. Transverse suture; 6. Scutum; 7. Pteropleuron; 8. Pteropleural setae; 9. Postalar callus; 10. Post alar wall; 11. Supra sqamal ridge; 12. Scutellum; 13. Post scutellum; 14. Anatergite; 15. Pleurotergite; 16. Posterioventral declivity of thorax; 17. Barette; 18. Hypopleuron; 19. Sternopleuron; 20. Sternopleural bristle; 21. Prostigmatic bristle; 22. Propleural bristle; 23. Propleuron.

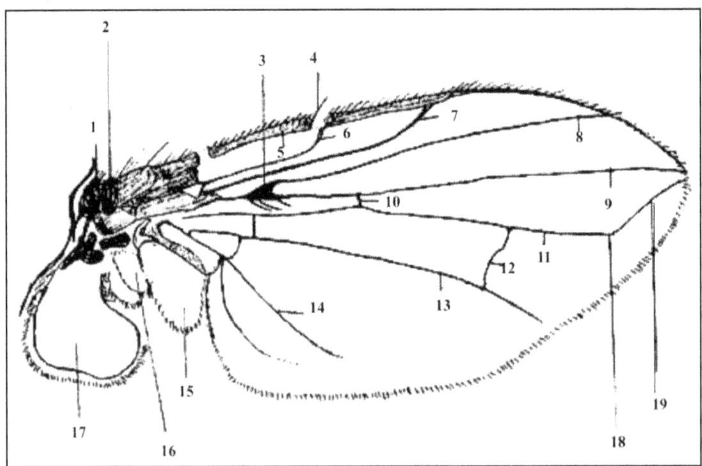

Figure 32: Morphological Features of Wing.

1. Tegula; 2. Basicosta 3. Base of r_{4+5}; 4. Costal spine; 5. Costa; 6. Costal section; 7. r_1; 8. r_{2+3}; 9. $r_{4+5;}$ 10. r-m; 11. m; 12. m-cu; 13. cu $_1$; 14. Anal wing; 15. Wing flap; 16. Wing scale; 17. Calypter; 18. Deflection of m; 19. Postangular vein.

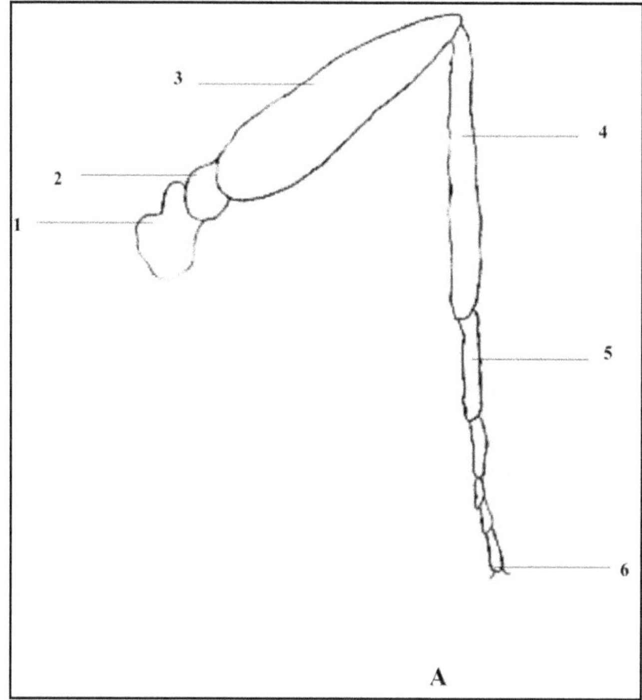

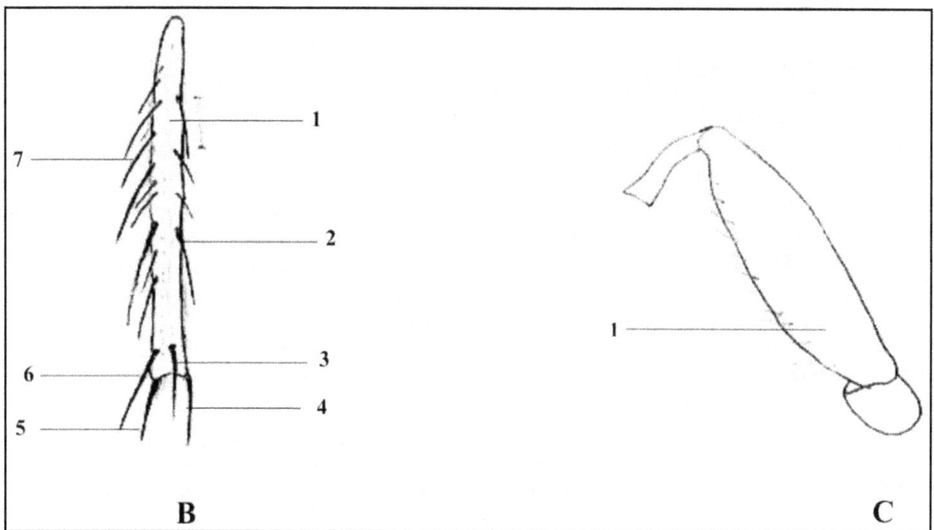

Figure 33: Morphological Features of Leg.

A. Leg: 1. Coxa; 2. Trochanter; 3. Femur; 4. Tibia; 5. Tarsus; 6. Claws.

B. Tibia: 1. Setulae; 2. Anterio ventral bristle; 3. Preapical anterioventral bristle; 4. Preapical posterioventral bristle; 5. Preapical posteriordorsal bristle; 6. Preapical anterodorsal bristle; 7. Anterodorsal bristle.

C. Femur.

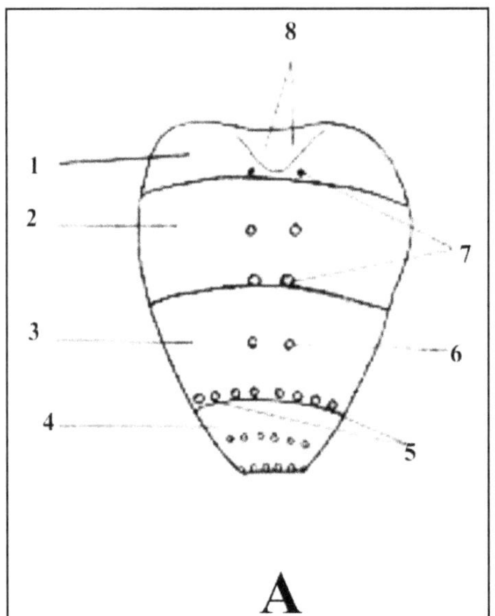

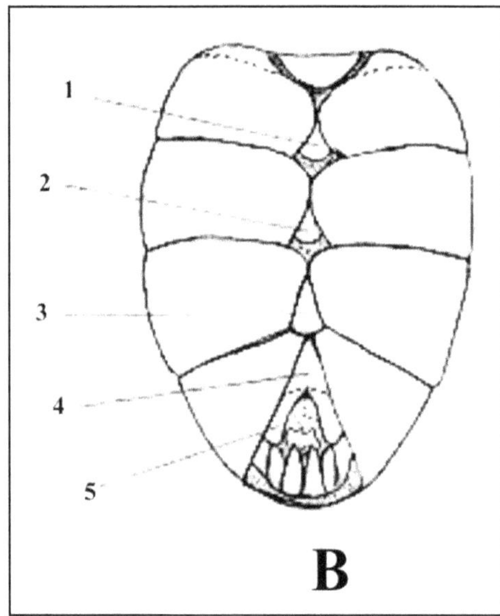

Figure 34: Morphological Features of Abdomen.

A: Dorsal view: 1. t_{1+2}; 2. t_3; 3. t_4; 4. t_5; 5. Marginal row; 6. Median discal bristle; 7. Marginal bristle; 8. Excavation of t_{1+2}.

B. Ventral view: 1. Sternite-2; 2. Sternite-3; Sternite-4; Sternite-5; 5. Lobe of sternite-5.

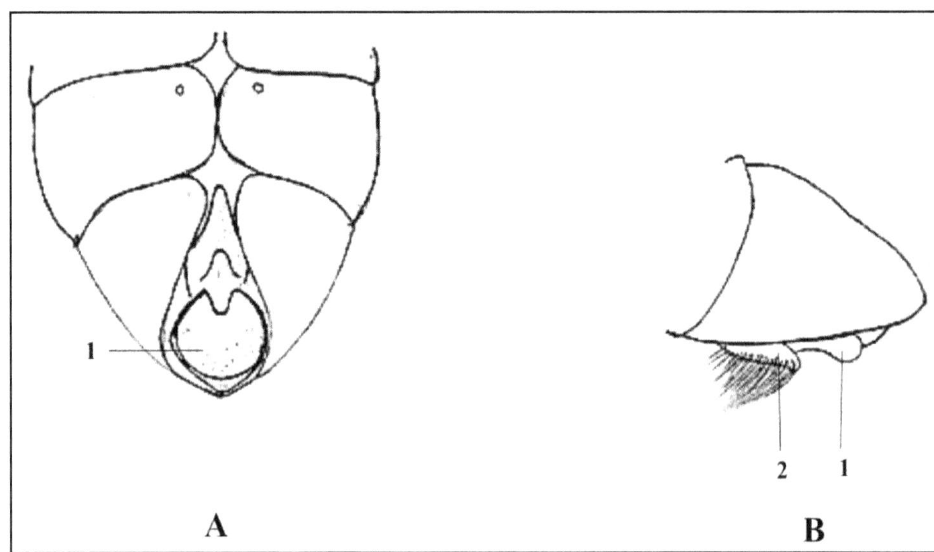

Figure 35: Male Genitalia.

A. Ventral View: 1. Epandrium.

B. Lateral View: 1. Epandrium; 2. Sternite-5.

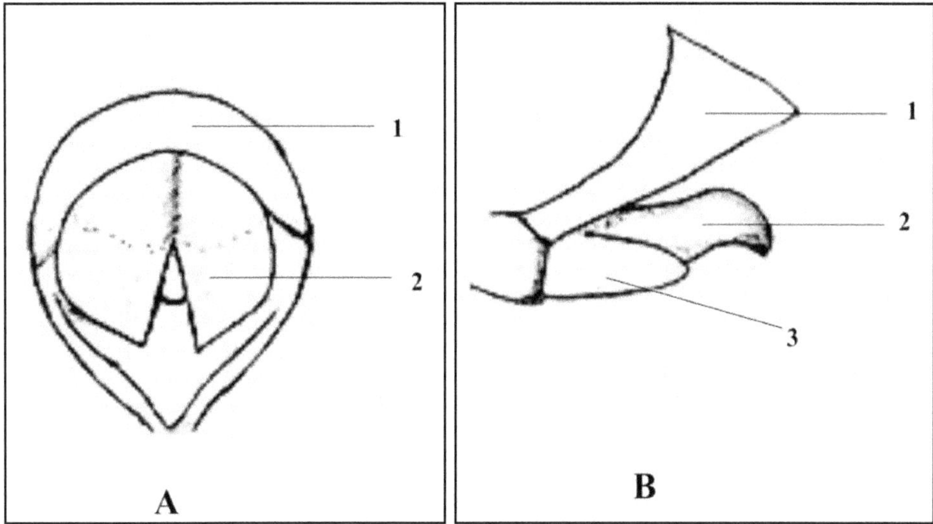

Figure 36: Female Genitalia.

A. Ventral View: 1. Tergite-6; 2. Tergite-7.

B. Lateral View: 1. Tergite-5; 2. Sternite-7; 3. Sternite-6.

Family Tachinidae is divided into four subfamilies *viz.*, Tachininae, Exoristinae, Phasiinae and Dexiinae. All these families are selected for studying the taxonomy.

Family Tachinidae

The Family Tachinidae is characterized by:

1. Subscutellum convexed.
2. Abdomen with spiracles situated in ventral part of tergites.
3. Prosternum hairy or bare.
4. Presence of well-developed ptilinal suture.
5. Exlusively parasitic on insect pests.

It contains four subfamilies namely, Tachininae, Exoristinae, Phasiinae and Dexiinae.

The Sub-family Tachininae (Macquartiinae) is characterized by

1. Prosternum with one/more black setulae.
2. Arista more or less bare.
3. Frontal setae descending below the base of antennae.
4. Jowls narrows.
5. Prealar setae shorter and finer than first post dc.

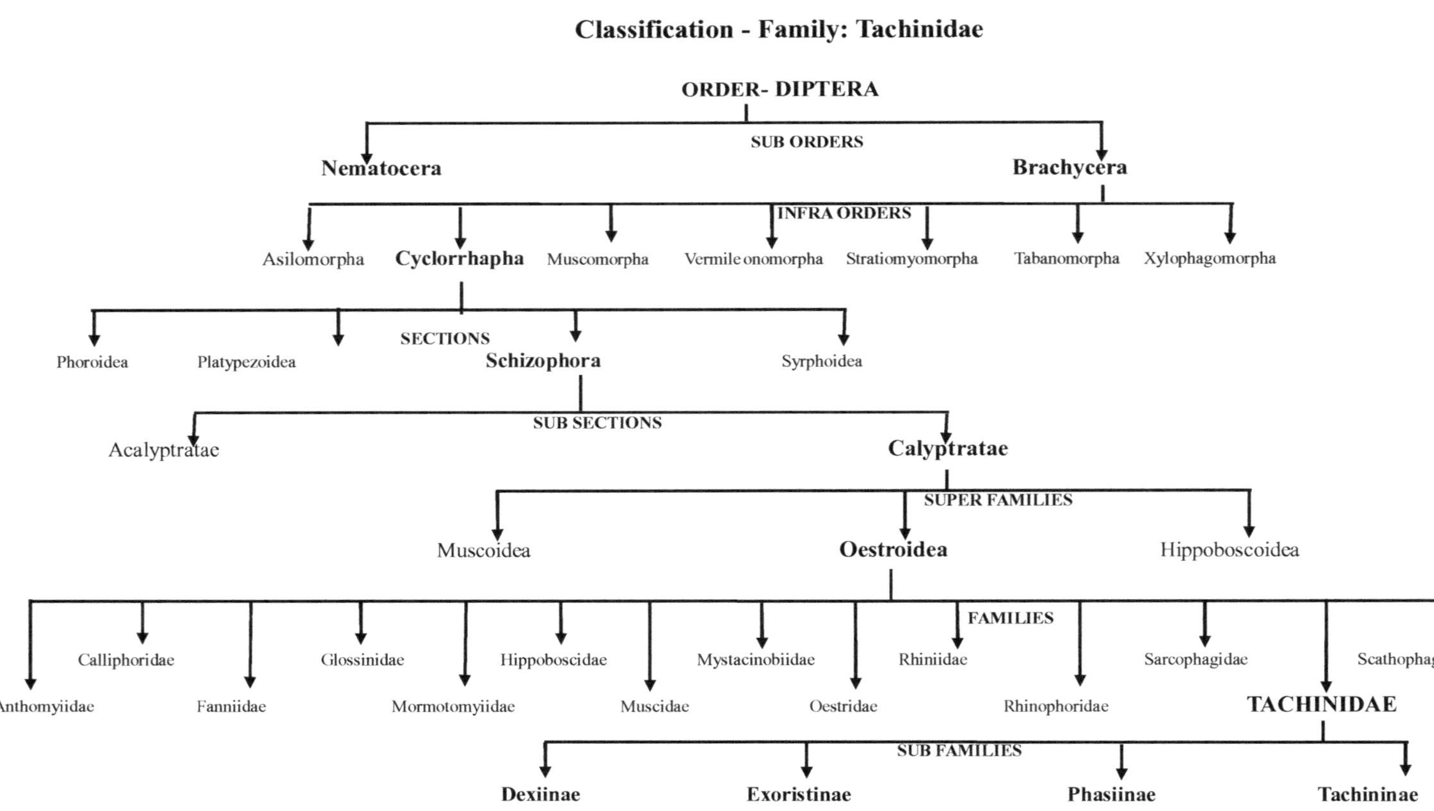
Classification - Family: Tachinidae
ORDER- DIPTERA
SUB ORDERS
Nematocera
Brachycera
INFRA ORDERS
Asilomorpha
Cyclorrhapha
Muscomorpha
Vermile onomorpha
Stratiomyomorpha
Tabanomorpha
Xylophagomorpha
SECTIONS
Phoroidea
Platypezoidea
Schizophora
Syrphoidea
SUB SECTIONS
Acalyptratae
Calyptratae
SUPER FAMILIES
Muscoidea
Oestroidea
Hippoboscoidea
FAMILIES
Anthomyiidae
Calliphoridae
Fanniidae
Glossinidae
Mormotomyiidae
Hippoboscidae
Muscidae
Mystacinobiidae
Oestridae
Rhiniidae
Rhinophoridae
Sarcophagidae
TACHINIDAE
Scathophagidae
SUB FAMILIES
Dexiinae
Exoristinae
Phasiinae
Tachininae

6. Basal excavation of first abdominal segment behind scutellum not reaches hind margin.
7. Preapical anteriodorsal seta on fore tibia as long as preapical dorsal seta.

Sub-family Exoristinae (Goninae) is characterized by

1. Prealar setae as long and strong as the first post dc or even stronger.
2. Basal excavation of first abdominal segment always reaching hind margin.
3. Arista bare.
4. Prosternum setose.
5. Frontal bristles usually descending on parafacials.
6. Preapical antero dorsal setae on fore tibia distinctly shorter than preapical dorsal setae.
7. Pteropleural setae clearly differentiated.

Sub-family Phasiinae is characterized by

1. Reclinate upper frontal setae normally absent or very seldom.
2. Never more than two posterior intra alars with wide separation.
3. Arista bare.
4. Base of abdomen excavated behind scutellum never reaches hind margin.
5. Prosternum bare.
6. Frontal bristles not descending on parafacials.
7. Presutural and anepimeral bristles absent or very short.
8. Facial ridges and parafacials bare.
9. Wing veins bare except base of r_{4+5}.

Sub-family Dexiinae (Proseninae) is characterized by

1. Arista bare/hairy/generally plumose.
2. Three posterior intra alars present.
3. Jowls (Lower cheeks) are wide.
4. Eyes small, never hairy.
5. Antennae short inserted below middle of eyes.
6. Frontal setae very seldom descending below base of antennae.
7. Prosternum bare or setulose.
8. Facial carina strongly elevated from the face so as to form a heavy ridge (Bulbous or flattened).
9. Gena very large in relation to eye.
10. Parafacials bare, pteropleural setae present.
11. Fore coxa bare on inner anterior surface.

Key to Subfamilies of the Family Tachinidae

1. Eyes hairy.....2
– Eyes bare.....3
2. Prosternum bare.....**Tachininae**
– Prosternum haired or setulose.....**Exoristinae**
3. Face raised into heavy fusiform facial carina.....**Dexiinae**
– Face not raised into heavy fusiform facial carina.....4
4. Prosternum haired or setose.....**Exoristinae**
– Prosternum bare.....5
5. Arista almost bare; facial ridge bare; reclinate orbital setae absent; 0-2 post intra alars; presutural intra alar setae absent; abdominal t $_{1+2}$ not excavate to its hind margin.....**Phasiinae**
– Forms without combinations given as above (5).....6
6. Arista long, plumose or setulose; facial ridge with few hairs; reclinate orbital setae present; 2-3 post intra alars; presutural intra alar setae present; abdominal t $_{1+2}$ usually excavate to its hind margin.....**Dexiinae**
– Forms without combinations given as above (6).....**Tachininae**

Sub-family Tachininae

Sub-family tachininae was erected by Van Embden (1960) under the name Macquartiinae. Later, Crosskey (1976) visulised this family as Tachininae. He included 22 tribes and 72 genera under this sub-family.

Key to Tribes of Sub-family Tachininae

1. Prosternum of thorax clearly inflated easily visible outside in profile.....Ormiini
– Prosternum of thorax normal not visible outside in profile.....2
2. Lower calypter very small.....Oxyphyllomyiini
– Lower calypter normal in size.....3
3. Hind coxa bare on posterodorsal surface.....4
– Hind coxa with small hairs on posterodorsal surface.....**Tachinini**
4. Thoracic surface with microrugose. no intra alars.....Germariochaetini
– Thoracic surface normal. 2 intra alars.....5
5. Ocellar setae reclinate. propleuron haired.....Campylochetini
– Ocellar setae proclinate. propleuron bare.....6

6. Parafacials with row of several strong downcurved setae ....Wagneriini
– Parafacials bare or but without strong setae..7
7. Fore coxa haired on its inner anterior surface 8
– Fore coxa bared on its inner anterior surface 11
8. Vein r_1 setose 9
– Vein r_1 bare 10
9. Arista plumose .... **Thelairini**
– Arista bare.. Voriini
10. Eyes haired. scutellum with four or more pairs marginal setae .... Nemoraeini
– Eyes bare. scutellum with three pairs marginal setae ....Micropthalimini
11. Supra alar region with one strong seta. One or two posterio intra alar setae ....Palpistomatini
– Supra alar region with two or three setae. Two or three posterio intra alar setae...12
12. Epistome prominent. Vibrissae inserted above the level of the epistomal margin 13
– Epistome not prominent. Vibrissae inserted on the level of the epistomal margin 16
13. Scutellum with two or three pairs of marginal setae. Humeral callus with two or three setae. Eyes bare .... Leskiini
– Scutellum with four or more pairs of marginal setae. Humeral callus with four or more setae. Eyes haired...14
14. Abdominal T1+2 excavate to its hind margin. Three postero intra alar setae 15
– Abdominal T1+2 not excavate to its hind margin. Two postero intra alar setae .... Parerigonini
15. Palpi completely developed.. Ernestiini
– Palpi very small or absent ....Linnaemyini
16. Eyes densely haired...**17**
– Eyes bare 18
17. Parafacials completely haired ....Macquartiini
– Parafacials almost bare... Ernestiini
18. Palpi absent. Face strongly excavate..Eloceriini
– Palpi present. Face weakly excavate...19

19. Abdomen round. Abdominal base and metacoxae very close together .... Glaurocarini
- Abdomen elongated fusifrom. Abdominal base and metacoxae widely separated 20

20. Parafacials haired .... **Phyllomyini**
- Parafacials bared 21

21. Abdomen more or less laterally compressed .... Minthoini
- Abdomen with without lateral compression .... **Thelairini**

Tribe Tachinini

Tribe tachini was raised by Mensil in 1966. According to Crosskey (1976), 10 genera have been reported from oriental region.

Key to Genera of Tribe Tachinini

1. Wing with first and third veins *i.e.* r_1 and r_{4+5} setulose. Parafacials with few strong setae .... *Tothilla* Cross.
- Wing with first and third veins *i.e.* r_1 and r_{4+5} totally bare. Parafacials without few strong setae 2

2. Propleuron bare. Palpi absent. 1+3 intra alar setae .... ***Peleteria*** R.D
- Propleuron haired. Palpi present. 1+2 intra alar setae 3

3. Hind coxa bare on posterodorsal surface .... *Schineria* Rond.
- Hind coxa with small hairs on posterodorsal surface 4

4. Abdominal t_{1+2} excavation without median marginal setae. Second costal sector haired ventrally .... *Mikia* Kow.
- Abdominal t_{1+2} excavation with median marginal setae. Second costal sector bare ventrally 5

5. Mesonotum, scutellum and abdomen metallic blue-violet .... *Chrysomikia* Mens.
- Non metallic without coloration 6

6. One post intra alar setae. One presututral dorsocentral setae. Humeral callus with three setae .... *Sericotachina* Town.
- Two post intra alar setae. Two or more presututral dorsocentral setae. Humeral callus with five setae 7

7. Abdominal tergite 3rd and 4th without discal setae. Two or more sternopleural setae 8
- Abdominal tergite 3rd and 4th with discal setae. One sternopleural setae .... *Eristaliomyia* Town.

8. Tarsi completely black .... *Nowckia* Wachtl
- Tarsi pale or yellow orange or red 9
9. Abdominal t $_{1+2}$ with one pair of median marginal setae.............. *Tachina* Meig.
- Abdominal t $_{1+2}$ without row of median marginal setae .... *Servillia* R. D

Genus – *Peleteria* Robineau-Desvoidy, 1830

The genus *Peleteria* is raised by Robineau-Desvoidy in 1830. According to Rayner and Raper (2001) 5 species of this genus have been reported from the world. The genus is characterized by,

1. Parafacials with several strong setae
2. Palpi absent
3. Propleuron bare
4. Ocellar setae missing

Key to Species of the Genus *Peleteria*

1. 2nd antennal segment yellow 2
- 2nd antennal segment black 4
2. Tergites 3rd and 4th with discal bristles............................ *popeli* Portsh.
- Tergites 3rd and 4th without discal bristles 3
3. Peristomes, cheeks and ventral side of tergite 2 with black hairs ......*ferina* Zett.
- Peristomes, cheeks and ventral side of tergite 2 with yellow hairs.................... 5
4. Each tergite covered with dusting at its posterior edge............................ *varia* F.
- Tergite 3rd, 4th and 5th are dusted and covers larger area of each tergite.. *tricolor sp.n*
5. Tergites without dusting. Abdomen ventrally with black longitudinal stripe.. *prompta* Meig.
- Tergites with dusting. Abdomen ventrally without black longitudinal stripe .... *rubescens* R.D

Peleteria tricolor sp.n.

(Plate 6, Figures 37 to 45)

Male (Figures 37-45)

Body 12.65mm in length, 5.09 mm in width; wing 9.33mm in length, 4.11mm in width; antenna 1.40 mm long; halter 1.51mm length, 0.49mm width.

Head (Figure 40)

1.73 mm in length, 2.89mm in width; vertex wide; inner vertical bristles strong, reclinate; outer vertical bristles strong, reclinate; ocelli present; ocellar bristles hair small proclinate; frontal vitta well developed; frontal bristles arising in single row, lowest bristle at level of middle of eye; fronto orbital plate with single row of setae; lateroclinate setae one pair; proclinate orbital setae with pair of two strong and one very small proclinate setae; face concave; lower facial margin concave; vibrissa strong, long; facial ridge concave; peristome and ventral side of tergite 2nd with yellow hairs; parafacial with stout bristles with 2 setulae; Cheek also shows additional white hairs with yellow hairs; gena narrow, with yellow hairs; genal dilation well developed; back of head slightly convex with pale yellow white hairs. **Eyes** 1.01mm in length, 2.49mm in width, bare, reddish brown. **Antenna** (Figure 40) scape very short, 0.09mm in length, 0.13mm in width; pedicel 0.88mm in length, 0.14mm in width, yellow; first flagellomere 0.43 mm in length, 0.13mm in width, concave, Third antennal segment is faint brownish in color. **Arista** bare. **Mouth** proboscis well developed; prementum short with labella; palpi absent.

Antennal Formula

S L/W=0.692, P L/W=6.285, F L/W=3.307, A= 3.428

Thorax (Figures 41 and 42)

4.89mm in length,4.09mm in width; humeral callus with reclinate 4 setae with few setulae; proepisternum with hairs; proepimeron with 3setae (2 postigmatic

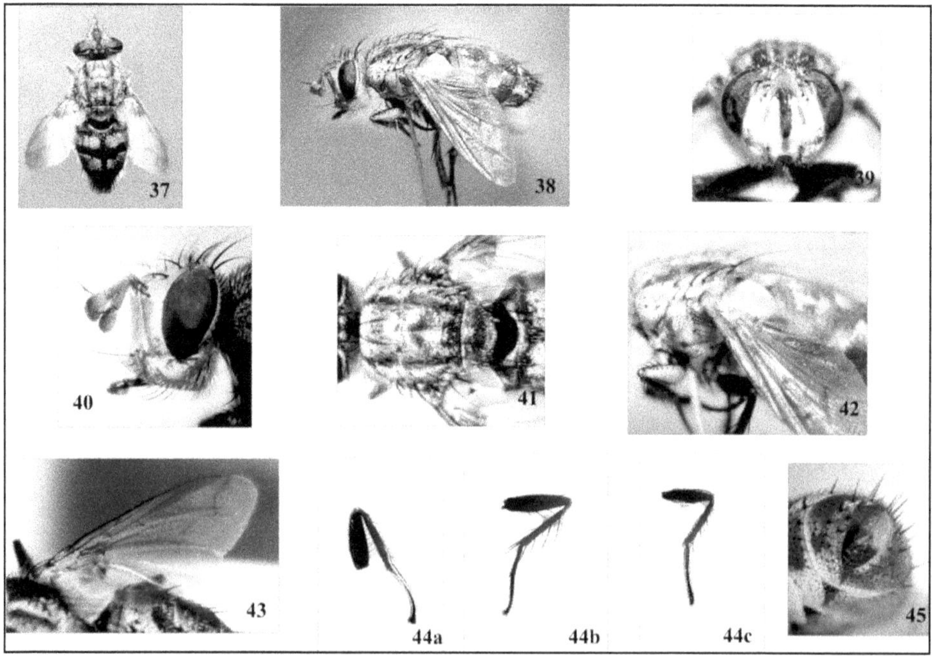

Plate 6: *Peleteria tricolor* sp.n. (Male) Dorsal view. Figure 38: Lateral View; Figure 39: Frontal View of Head; Figure 40: Antennae; Figure 41: Dorsal View of Thorax; Figure 42: Lateral View of Thorax; Figure 43: Fore Wing; Figure 44a: Fore Leg; Figure 44b: Mid Leg; Figure 44c: Hind Leg; Figure 45: Genitalia.

setae + 1 propleural setae). **Scutum** 3.65 mm in length, 3.98mm in width, with 4 longitudinal strips; acrostichal bristles (presutural 3+ postsutural 3); dorsal central bristles (presutural 3+ postsutural 4); notopleuron with two reclinate setae; postalar callus with two setae. **Scutellum** 1.24 mm in length, 2.16mm in width, tip of scutellum faint brownish in color; basal scutellar bristles slightly curved; lateral scutellar bristle parallel; subapical scutellar bristles long, divergent; apical scutellar bristles crossed; subscutellum convex; katepisternum with 3 setae; katepimeron bare; anepimeron with one setae; meron with row of setae; anatergite with small hairs; anterior thoracic spiracle narrow and closed by fringes of hairs.

Wing (Figure 43)

9.33mm in length, 4.11mm in width, transparent; lower calypter well developed, hind margin less contiguous to margin of scutellum; second costal section bare, pale yellow patch; base of costa with small hairs; costal bristles not differentiated than costigeal setae; fourth costal section as long as 6th costal section; r_1 bare; cu a_1 bare; bend of m distinct; wing cell r_{4+5} open; cross vein r-m present; cross vein d m-cu present; anal vein not reaches towards hind margin of wing; halter reddish yellow, 1.51mm length, 0.49mm width.

Leg

Fore and mid femora with two rows of spines at ventrally; **Forelegs** (Figure 44a) femora 3.21mm; tibia 3.48mm and tarsus 3.79mm; foreleg with one preapical posteriodorsal seta, one long preapical anterioventral setae, anteriodorsal setae comb like. **Midlegs** (Figure 44b)- femora 3.15mm; tibia 3.34mm and tarsus 3.76mm, mid tibia with one preapical anteriodorsal setae, one preapical posteriodorsal seta, one preapical anterioventral setae, one preapical posterioventral seta, 4 anteriodorsal setae long, spiny. **Hind legs** (Figure 44c)- femora 3.19mm; tibia 3.49mm and tarsus 3.81mm; hind tibia with one preapical posteriodorsal seta, one preapical anterioventral setae, one long, strong posterioventral seta; hind coxa with two setae.

Abdomen (Figures 37 and 38)

6.03 mm in length, 4.15mm in width; abdominal tergite 5th as long as 4th tergite; mid dorsal depression 1+2 syntergite reaches up to 2 tergite margin; discal bristles present; Tergite 3rd, 4th and 5th are dusted and covers larger area of each tergite. Abdominal sternites overlapped by ventral edges of tergites.

Genitalia (Figure 45)

Terminalia retracted deeply within abdominal tergite 5th; cercus, surstylus attaches to epandrium deeply situated in tergite 5th.

Colour

Black: Fore Tibia, Mid Tibia, Hind Tibia,

Brown: Frontal vita, Antenna, Palpus,

Golden Yellow: Parafrontal, vertex

Silvery white: Head, Face, Parafacial, Thorax,

Reddish brown: Eyes, Abdomen

Host- Unknown

Host Plant- *Symphyotrichum* sp.

Holotype- Male, India, Maharashtra, Coll. 13-X-2013, Mahabaleshwar, A.S. Desai; legs on slides, labeled as above.

Paratype- 3 Males, Sex ratio (M: F) 3:0, July 2011- February 2015; legs on slides, rest part pinned, preserved in insect box.

Etymology- The species name is due to three colours on body parts.

Distributional Record

India, Maharashtra, ♂ 1 Jath 09-XI- 2013; ♂2 Mahabaleshwar 13-X-2013.

Remarks

According to key of Crosskey, 1976 and Rayner and Raper, 2001, this species runs close to *Peleteria varia* (Fabricius, 1794) by having following characters,

1. Peristome and ventral side of tergite 2nd with yellow hairs.
2. Palpi absent.

However, it differs from the above species by following distinct characters,

1. Tergite 3rd, 4th and 5th are dusted and covers larger area of each tergite.
2. Cheek also shows additional white hairs with yellow hairs.
3. Tip of scutellum faint brownish in color.
4. Third antennal segment is faint brownish in color.
5. **Antennal formula:** S L/W=0.692, P L/W=6.285, F L/W=3.307, A= 3.428

Tribe Thelairini

Tribe Thelairini was raised by Mensil (1973) and supported by Crosskey (1975). According to Crosskey (1976), 10 genera have been reported from oriental region.

Key to Genera of Tribe Thelairini

1. Scutellum with three or more pairs of marginal setae 2
– Scutellum with two pairs of marginal setae 9
2. Scutellum with four or more pairs of marginal setae. Apical scutellar setae divergent.. *Xanthopteromyia* Town.
– Scutellum with total three pairs of marginal setae. Apical scutellar setae crossed and horizontal 3
3. Head with a series of 5 to 7 pairs of proclinate orbital setae....... *Halydaia* Egger
– Head without proclinate orbital setae 4

4. Abdomen with strong median discal setae on t_3 and t_4 5
– Abdomen without strong median discal setae on t_3 and t_4 7
5. Abdomen long, slender and fusiform, t_{1+2} not excavate. Subapical scutellar setae not widely separated .... *Polygastropteryx* Mens.
– Abdomen subovate, t_{1+2} excavate. Subapical scutellar setae widely separated 6
6. Fore coxa with short fine hairs on the whole inner anterior surface. 3 post intra alar setae. Arista short plumose...................... ***Thelaira*** R. D
– Fore coxabare. 2 post intra alar setae. Arista pubescent .... *Prosheliomyia* B.B
7. Mid tibia with 2 anterodorsal setae. 3 postdorsocentral setae. Notopleuron with pale yellow hairs .... *Allothelaira* Vill.
– Mid tibia with 1 anterodorsal setae. 4 postdorsocentral setae. Notopleuron completely bare 8
8. 3 sternopleural setae. Scutellum with very much strong subapical setae .... *Actinochaetopteryx* Town.
– 2 sternopleural setae. Scutellum with weak subapical setae.. *Thryptodexia* Mall.
9. Head profile conspicuously subtriangular. Antennae very large. Arista bare. 2 sternopleural setae .... *Zambesa* Walk.
– Head profile not at all subtriangular. Antennae small. Arista plumose. 3 sternopleural setae.. *Zambesa* Walk.

Genus – *Thelaira* Robineau-Desvoidy, 1830

The genus *Thelaira* is raised by Robineau-Desvoidy, 1830. According to Rayner and Raper (2001) 3 species of genus *Thelaira* have been reported from the world. The genus is characterized by:

1. Three post intra alar setae.
2. Pteropleural set absent.
3. Fore coxa with short fine uniform hairs on it anterior surface.

Key to Species of the Genus *Thelaria*

1. Middle tibia with 2 anterodorsal setae. Calyptrae white, the edge usually yellowish .... ***nigripes*** F.
– Middle tibia with 3-4 anterodorsal setae 2
2. Outer vertical bristle weak. Calyptrae yellowish............................ *solivaga* Harr.
– Outer vertical bristle weak. Calyptrae yellowish..........................*leucozona* Meig.

***Thelaria nigripes* (Fabricus, 1794)**

(Plate 7, Figures 46 to 63)

Male (Figures 46-54)

Body 12.42 mm in length, 4.95 mm in width; wing 10.86mm in length, 4.26mm in width; antenna 1.22 mm long; halter 1.09mm in length, 0.41mm in width.

Head (Figure 48)

0.92mm in length, 3.09mm in width; vertex narrow; inner vertical bristles strong, single pair; outer vertical bristle absent; ocelli present; ocellar bristles single, strong; frontal vitta well developed, dark black; frons narrow; frontal bristles inwardly proclinate with; fronto orbital plate silvery white; reclinate upper orbital bristles absent; proclinate orbital bristles absent; face silvery-white, shorter than frons; lower facial margin short; vibrissa present on a level of epistomal margin; facial ridge not clearly seen through lateral view; parafacial silvery whitish; gena narrow, visible in lateral view; genal dilation well developed; back of head with few hairs, fine and short, faint yellow. **Eyes** 0.90mm in length, 2.88mm in width, bare, well developed. **Antenna**(Figure 49) scape 0.15mm in length, 0.11mm in width, short; pedicel short 0.20mm in length, 0.15mm in width; first flagellomere 0.88mm in length, 0.14mm in width, concave. **Arista** plumose type, with longer hairs. **Mouth** proboscis well developed; prementum shorter; palpus well developed, slightly swollen apically.

Antennal Formula

S L/W=1.363, P L/W=1.34, F L/W=6.285, A=2.996

Thorax (Figures 50 and 51)

4.62mm in length, 3.22mm in width, with 4 strips present before suture; proepisternum with 4 setae; proepimeron with 2 dorsally directed setae; prosternum with short hairs. **Scutum** 3.08mm in length, 3.14mm in width, with four narrow longitudinal stripes and acrostichal bristles (presutural 3+ postsutural 3); dorsal central bristles (presutural 3+ postsutural 3); notopleuron with strong bristles; post sutural intra alars with 3 setae; postalar callus 2 setae. **Scutellum** 1.54mm in length, 1.92mm in width; basal scutellar bristles shorter than subapical setae, lateral scutellar bristles absent; sub apical scutellar bristles stronger, widely separated; apical scutellar bristles crossed; discal setae with pair of setae; subscutellum convexed; katepisternum with 2 sternopleural setae; anepimeron 4 strong row of setae with hairs; katepimeron bare; meron with vertical row of small bristles; postmetacoxal area present; anatergite bare; anterior thoracic spiracle narrow.

Wing (Figure 52)

10.86mm in length, 4.26mm in width; lower calypter, white with yellowish edge, well developed; second with fine hairs; base of costa without costigial bristles; costal spine absent; fourth coastal bristles open; r_1 bare; cu a_1 rarely setose; bend of m distinct; wing cell r_{4+5} open, with small hairs; cross vein r-m present; cross vein d m-cu present; anal vein not reaching hind margin of wing. **Halter** dark yellow, 1.09mm length, 0.41mm width.

Plate 7. Figure 46: *Thelaria nigripes* Fabricus (Male) Dorsal view; Figure 47: Lateral View; Figure 48: Frontal View of Head; Figure 49: Antenna; Figure 50: Dorsal View of Thorax; Figure 51: Lateral View of Thorax; Figure 52: Fore Wing; Figure 53a: Fore Leg; Figure 53b: Mid Leg; Figure 53c: Hind Leg; Figure 54: Genitalia.

Leg

Fore coxa predominantly covered with setulae, **Forelegs** (Figure 53a)femora 4.92mm; tibia 4.26mm; tarsus 3.32mm; fore tibia with preapical anteriodorsal; fore tarsus with compressed 2 setae. **Midlegs** (Figure 53b) femora 4.64mm; tibia 4.12mm; tarsus 3.29mm, mid tibia with two anteriodorsal setae. **Hind legs** (Figure 53c) femora 4.36mm; tibia 4.19mm; tarsus 3.28mm; hind tibia with one preapical posteroventral seta, 2 anterodorsal bristles; hind coxa with many setae.

Abdomen (Figures 46 and 47)

7.14mm in length, 4.09mm in width, ovate; abdominal tergite 5th is as long as abdominal tergite 4th; mid dorsal depression syntergite 1+2 reaching towards margin; discal bristles present; abdominal sternites overlapped by ventral edges of tergites; mid dorsal hairs of tergite 3 and 4 as long as marginal bristles.

Genitalia (Figure 54)

Male terminalia retracted within abdominal tergite 5; tergite 6th less reduced, joined with 7th tergite; cercus and syrstylus present below syntergite 7+8.

Colour

Black: Thorax, fore tibia, mid tibia, hind tibia.

Brown: Antenna.

Dark brown: Frontal vita.

Yellow: Abdomen.

Silvery white: Head, face, parafacial.

Reddish brown: Eyes, palpus.

Female (Figures 55-63)

Body 10.36 mm long, 3.22 mm broad; wing 8.54mm in length, 3.72mm in width; antenna 1.25 mm long; halter 1.09mm in length, 0.41mm in width.

Head (Figure 57)

1.33mm in length, 3.15mm in width; vertex narrow; inner vertical bristles strong, reclinate; outer vertical bristle absent; ocelli present; ocellar bristles present, proclinate single pair with small hairs; frontal vitta black; frontal bristles inwardly proclinate; fronto orbital plate silvery white; lateroclinate upper orbital bristles single pair; proclinate orbital bristles one pair; face silvery white; lower facial margin partially seen in lateral view; vibrissa strong well developed crossing to each other; facial ridge concave; parafacial silvery white; gena narrow; genal dilation with well developed three strong setae; back of head convex, silvery white. **Eyes** 1.3mm in length, 2.48mm in width, bare, reddish brown. **Antenna** (Figure 58) scape 0.14mm in length, 0.11mm in width; pedicel 0.29mm in length, 0.14mm width, silvery white; first flagellomere 0.82mm in length, 0.14mm in width. **Arista** plumose type, arises from base of first flagellomere; **Mouth** proboscis well developed, but not clearly seen in lateral view; prementum shorter; palpus slightly swollen apically with some hairs.

Antennal Formula

S L/W=1.272, P L/W=2.071,F L/W=5.857, A=3.067

Thorax (Figures 59 and 60)

4.06mm in length, 2.74mm in width; proepisternum with 2 strong setae and small setulae; proepimeron with 1 propleural setae; prosternum with hairs. **Scutum** 2.66mm in length, 2.70mm in width; acrostichal bristles (presutural 3+ postsutural 3); dorsal central bristles (presutural 2+ postsutural 3); notopleuron with 2 strong setae; postalar with 2 strong setae. **Scutellum** 1.04mm in length, 1.19mm in width; basal scutellar bristles with one strong; one lateral scutellar bristles stronger and longer than others. **Subscutellum** convex, bare; katepisternum 2 strong; anepimeron bare; katepimeron bare; meron few reclinate row of setulae; postmetacoxal area with setulae; anatergite bare; anterior thoracic spiracle mesopleuron shows backward projecting row of setae.

Wing (Figure 61)

8.54mm in length, 3.72mm in width, lower calypter inner margin more or less contiguous to lateral margin of scutellum; second costal section cs_2; base of costa without strong long setae, minute hairs are present; costal bristles absent; fourth coastal bristles longer than 6th costal section; r_1 with small setulae externally; cu a_1 bare; bend of m distinct; wing cell r_{4+5} at based end small setulae present dorsally;

cross vein r-m present; cross vein d m-cu present; anal vein reaches towards hind margin of wing. **Halter** reddish, 1.29mm in length, 0.38mm in width.

Leg

Fore and mid femora with two rows of spines ventrally. **Fore Leg** (Figure 62a) femora 3.86mm, tibia 3.24mm, tarsus 2.84mm, fore tibia with one preapical anterioventral setae. **Mid leg** (Figure 62b) femora 3.52mm, tibia 3.01mm, tarsus 2.45mm, mid tibia with two anterior dorsal bristles, one preapical anteriodorsal setae, one preapical posteriodorsal setae. **Hind legs** (Figure 62c) femora 3.77mm, tibia 3.37mm, tarsus 2.71mm, hind tibia one preapical anteriodorsal setae, one preapical posteriodorsal setae; hind coxa with some setae.

Abdomen (Figures 55 and 56)

5.28mm in length, 3.02mm in width; abdominal tergite 5th fused small; mid dorsal depression not fuses to margin of second segment; discal bristles with 2 rows. Abdominal sternites overlapped by ventral edges of tergites.

Genitalia (Figure 63)

Terminalia retracted within abdominal tergite 5; tergite 6th less reduced, joining 7th segment of tergite; piercer present below syntergite 7+8.

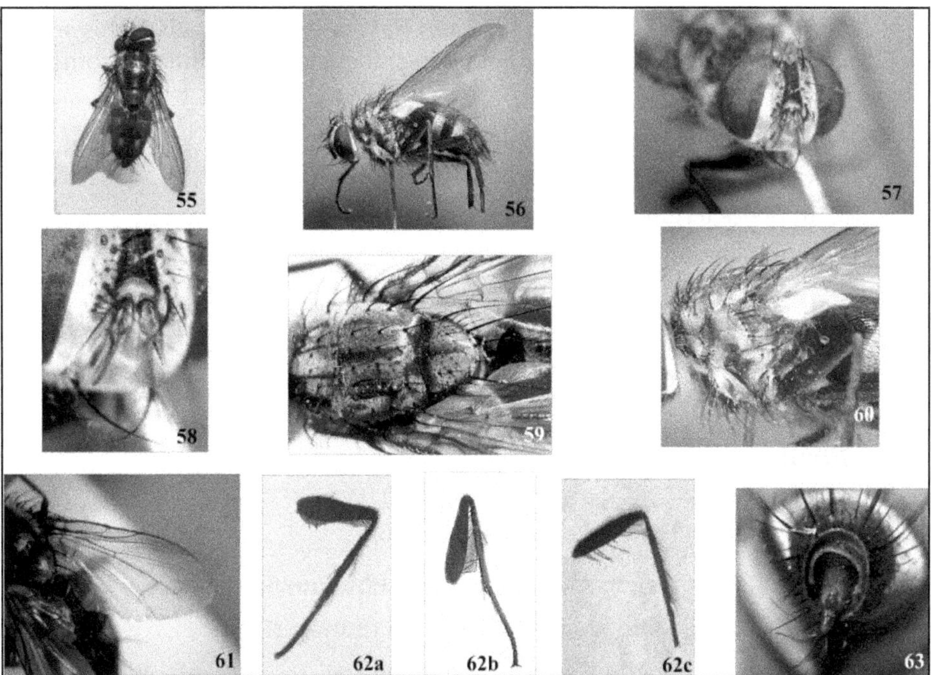

Plate 8. Figure 55: *Thelaria nigripes* Fabricus (Female) Dorsal view; Figure 56: Lateral View; Figure 57: Frontal View of Head; Figure 58: Antenna; Figure 59: Dorsal View of Thorax; Figure 60: Lateral View of Thorax; Figure 61: Fore Wing; Figure 62a: Fore Leg; Figure 62b: Mid Leg; Figure 62c: Hind Leg; Figure 63: Genitalia.

Colour

Black: Thorax, abdomen, fore tibia, mid tibia, hind tibia.

Brown: Head, palpus.

Dark brown: Frontal vita.

Silvery white: Face, parafacial, antenna.

Reddish brown: Eyes.

Host- Unknown.

Host Plant- *Solidago canadensis, Rorippa indica.*

Paratype- Sex ratio (M: F) 15:4, July 2011- February 2015; legs on slides, rest parts pinned, preserved in insect box.

Distributional Record

India, Maharashtra, ♂1 ♀1 Kolhapur 16- VIII-2012; ♂3 ♀1 Panhala 15-VIII- 2013; ♂2 Islampur 08-XI-2013; ♂ 1 Kankwali 15-VIII- 2013; ♂ 2 ♀1 Chandgad 19-VIII- 2013; ♂1 Radhanagari 23-VIII- 2013; ♂ 2 Khanapur 25-VIII- 2013; ♂ 1 Koyana 14-XI- 2013; ♂ 1 Shahuwadi 20-X- 2013; ♂ 1♀1 Patan 01-X- 2014.

Remarks

According to key of Crosskey, 1976 and Rayner and Raper, 2001, this species is *Thelaira nigripes* (Fabricus, 1794). Following characters are similar according to above key,

1. Middle tibia with 2 anterodorsal setae.
2. Calyptrae white, the edge usually yellowish.
3. In males tergite abdomen appears dark.
4. In females tergite 5th undusted.

Additional characters are observed as follows:

1. In males vibrissae are comparatively smaller than females.
2. **Antennal formula:** S L/W=1.272, P L/W=2.071,F L/W=5.857, A=3.067

Tribe Phyllomyini

Tribe Phyllomyini was raised by Mensil (1973). According to Crosskey (1976), 3 genera have been reported from oriental region.

Key to Genera of Tribe Phyllomyini

1. Parafacials haired. Fore tibia with 2 postero ventral setae 2
- Parafacials bare. Fore tibia with 1 postero ventral setae .... *Metopomintho* Town.
2. Abdominal tergite 3rd, 4th and 5th each with one pair of erect median discal setae. Lower calypter small and widely separated from scutellum. 2 or 3 postdorsocentral setae. Abdomen large, fusiform .... ***Phyllomya*** R.D

– Abdominal tergite 3rd, 4th and 5th each with irregular row of long erect discal setae. Lower calypter large and close to scutellum. 3 post-dorsocentral setae. Abdomen not distinctly fusiform............. *Gibsonomyia* Curr.

Genus – *Phyllomya* Robineau-Desvoidy, 1830

The genus was raised by *Phyllomya* Robineau-Desvoidy,in 1830. From various parts of the world more than 3 species of genus have been reported. The genus is characterized as follows:

1. Each abdominal tergite 3rd, 4th and 5th always with one pair of erect median discal setae on the dorsum.
2. Lower calypter small, separated from the scutellum.
3. Thorax with 2 to 3 post dorsocentral setae.
4. Abdomen long and fusiform.

Key to Species of the Genus *Phyllomya*

1. Bend of M distinct with small projection of M_2 on wing. Vibrissae weak not differentiated from hairs on vibrissal angle... ***panhalensis*** sp. n

– Bend of M distinct without small projection of M_2 on wing. Vibrissae clearly differentiated from hairs on vibrissal angle 2

2. Three postdorsocentral setae. Abdominal t_{1+2} without discal setae. Hind tibia without preapical posteriodorsal setae .... *gibsonomyiodes* Brun.

– Two postdorsocentral setae. Abdominal t_{1+2} with discal setae. Hind tibia with preapical posteriodorsal setae ....*elegans* vill.

Phyllomya panhalensis sp.n.

(Plate 9, Figures 64 to 72)

Male (Figures 64-72)

8.5mm in length, 0.29mm in width; head 2.13 mm in length, 0.76mm in width; thorax 2.92mm; abdomen 5.1mm; wing 8.52mm in length, 2.84mm in width; halter 1.53mm in length, 0.53mm in width.

Head (Figure 66)

2.13 mm in length, 0.76mm in width; vertex narrow; inner vertical bristles with single reclinate pair; outer vertical bristle single pair, proclinate orbital setae absent; ocelli present; ocellar bristles long; frontal vita dark brown, narrow; frontal bristles single row medioclinate; face concave silvery white; lower facial margin protruded forward; vibrissa small weak; facial ridge slightly concave; parafacial silvery white; gena with small hairs, gena and parafacials appearing silvery but not reddish brown; genal dilation weakly developed; cheeks with hairs; back of head convex with few hairs. **Eyes** 0.42mm in length, 1.29mm in width; bare. **Antenna**

(Figure 67) silvery-light brown with scape short 0. 07mm in length, 0.11mm in width; pedicel short 0.09mm, 0.148mm in width with few hairs, first flagellomere 0.62mm in length, 0.14mm in width, antennal third segment more than thrice as long as second segment. **Arista** plumose type. **Mouth** proboscis well developed; prementum short with labella; palpus filiform parallel sided, brownish.

Antennal Formula

S L/W=0.636, P L/W=0.61, F L/W=4.428, A=1.891.

Thorax (Figures 68 and 69)

2.92mm in length, 2.03mm in width; black with thin silvery pollinosity most evident on pleural regions; humeral callus three setae dorsally projected; proepisternum bare; proepimeron pair of setae present. **Scutum** 1.92 in length, 2.01mm in width; double dark bands; acrostichal bristles strong, reclinate (presutural 3+ postsutural 3); dorsal central bristles strong; reclinate (presutural 3+ postsutural 3); notopleuron with 2 setae; postalar callus with 2 setae. **Scutellum** 1mm in length, 1.02mm in width; basal scutellar bristles straight; subapical scutellar bristles very long than other, strong. **Subscutellum** convex; katepisternum with 2 strong setae; anepimeron with hairs; katepimeron bare; anatergite bare; anterior thoracic spiracle narrow, with closed fringes of hairs.

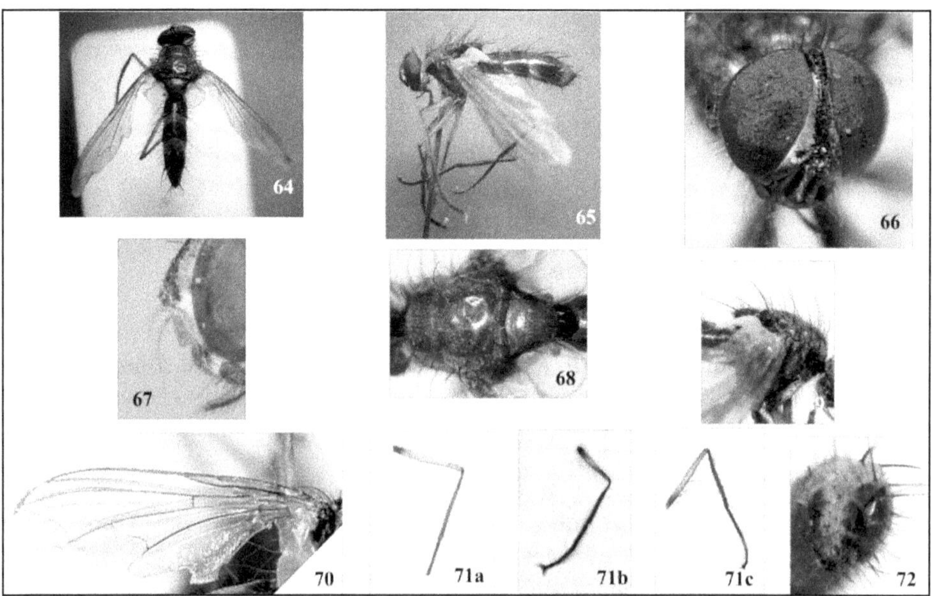

Plate 9. Figure 64: *Phyllomya panhalaensis* sp.n. (Male) Dorsal view; Figure 65: Lateral View; Figure 66: Frontal View of Head; Figure 67: Antennae; Figure 68: Dorsal View of Thorax; Figure 69: Lateral View of Thorax; Figure 70: Fore Wing; Figure 71a: Fore Leg; Figure 71b: Mid Leg; Figure 71c: Hind Leg; Figure 72: Genitalia.

Wing (Figure 70)

8.52mm in length, 2.36mm in width transparent with very faint brownish tinge to naked eye; lower calypter small slightly separated from scutellum; second costal section bare; base of costa with long costal bristle; costigial setae small; fourth costal section longer than 6th costal section; r_1 bare; cu a_1 bare; bend of m distinct and shows small projection of m_2; wing cell r_{4+5} open; cross vein r-m present; cross vein d m-cu present; anal vein not reaching towards hind margin of wing. **Halter** reddish yellow, 1.53mm in length, 0.53mm in width.

Leg

Fore and mid femora with row of spines ventrally, Legs brownish black with yellow Femur. **Forelegs** (Figure 71a) femora 3.73mm; tibia 3.28mm; tarsus 3.44mm; fore tibia yellow, with 1 preapical anterodorsal seta, fore tarsus brown. **Mid legs** (Figure 71b) femora 3.66mm; tibia 3.15mm; tarsus 3.37mm;fore tibia mid tibia yellow 1 preapical anteriodorsal setae present. **Hind legs** (Figure 71c) femora 3.69mm; tibia 3.21mm; tarsus 3.39mm;fore tibia hind tibia yellow,1 preapical posteroventral seta present, 1 anteriodorsal bristles present; hind coxa with 1 short setae.

Abdomen (Figures 64 and 65)

5.1mm in length, 1.54mm in width, long, fusiform, black with narrow white bands on anterior edge of tergites; abdominal tergite 3rd, 4th, 5th with one pair of erect median discal setae. abdominal tergite 5th as long as about 4th tergite; mid dorsal depression on syntergite 1+2 not reaching of that segment; tergite 5th tapered at end; discal bristles present abdominal sternite overlapped by ventral edges of tergites; spiracle remains on side of tergite 7.

Genitalia (Figure 72)

Terminalia retracted within abdominal tergite 5; tergite 6 less reduced, joining segment 7, abdominal tergite 7th and 8th joined to form syntergite 7+8. cercus and surstylus present below syntergite 7+8.

Colour

Black: Head, thorax, abdomen.

Brown: Antenna, palpus.

Dark brown: Frontal vita.

Yellow: Fore tibia, mid tibia, hind tibia.

Silvery white: Face, parafacial.

Reddish brown: Eyes.

Host- Unknown.

Host Plant- *Caltropis* sp.

Holotype- Male, India, Maharashtra, Coll. 8-VIII-2013, Panhala, A.S. Desai; legs on slides, labeled as above.

Paratype- 2 Males, Sex ratio (M: F) 2:0, July 2011- February 2015; legs on slides, rest part pinned, preserved in insect box.

Etymology- The species name is described from the Panhala region of Kolhapur, India where the type material has been collected.

Distributional Record

India, Maharashtra, ♂ 1 Panhala 13-VIII- 2013; ♂1 Shahuwadi 08-XI-2013.

Remarks

According to key of Crosskey, 1976, this species runs close to *Phyllomya gibsonomyides* by having following characters:

1. Genal dilation weakly developed.
2. Ocellar bristles long.
3. Palpus filiform- parallel sided, brownish.
4. Thorax black with thin silvery pollinosity most evident on pleural regions.
5. Wings transparent with very faint brownish tinge to naked eye.
6. Katepisternum with 2 sternopleural setae.

However, it differs from the above species by following distinct characters:

1. Gena and parafacials appearing silvery but not reddish brown.
2. Antenna silvery-light brown and antennal third segment more than thrice as long as second segment.
3. Vibrissae weak not differentiated from hairs on vibrissal angle.
4. Bend of M distinct and shows small projection of M_2 on wing.
5. Legs brownish black with yellow Femur.
6. **Antennal formula:** S L/W=0.636, P L/W=0.61, F L/W=4.428, A=1.891.

Sub-family–Exoristinae

Sub-family Exoristinae was erected by Crosskey (1973), first time he reported it as Goniinae but, recently it is known as Exoristinae. According to Crosskey (1976), 12 tribes and 126 genera have been reported from oriental region.

Key to Tribes of Sub-family Tachininae

1. Pre alar setae weak and shorter than the first post intra alar setae......................2
– Pre alar setae strong and longer than the first post intra alar setae 7
2. Basal node of r $_{4+5}$ with a single setula. 2-3 pairs postdorsocentral setae. Eyes bare .... Acemyini
– Basal node of r $_{4+5}$ with a two or more setula. 3-4 postdorsocentral setae. Eyes bare or hairy 3
3. Subapical scutellar setae crossing or converging..4
– Subapical scutellar setae subparallel or diverging from each other 5

4. Vein r_{4+5} with series uniform setulae. Excavation of abdominal t_{1+2} extending beyond middle tergite .... Siphonini
- Vein r_{4+5} with single strong setula. Excavation of abdominal t_{1+2} extending virtually to hind margin of tergite .... Neaerini
5. Bend of m on wing without trace of m_2. Subapical scutellar setae widely divergent .... **Blodeliini**
- Bend of m on wing with trace of m_2. Subapical scutellar setae subparallel or slightly divergent 6
6. Bend of m on wing with trace of m_2 or at least darkened fold. Eyes bare or haired .... **Exoristini**
- Bend of m on wing without trace of m_2 or at least darkened fold. Eyes haired .... Ethillini
7. Ocellar setae reclinate .... Goniini
- Ocellar setae proclinate or absent 8
8. Eyes very large and gena very much reduced 9
- Eyes comparatively small and genal depth as like width of third antennal segment 10
9. Humeral callus with five setae. Parafacials haired. Eyes densely hairy .... **Winthemiini**
- Humeral callus with not more than four setae. Parafacials bare. Eyes bare or hairy .... **Carceliini**
10. Humeral callus with five setae. Eyes densely hairy .... **Winthemiini**
- Humeral callus with fewer than 4 setae. Eyes densely hairy or bare 11
11. Vibrissae inserted at a level above epistomal margin. 3+4 dorsocentral setae. Inner posterior angle of lower calypter well developed .... **Sturmiini**
- Vibrissae inserted on the level epistomal margin. Various complements of dorsocentral setae. Inner posterior angle of lower calypter rounded .... **Eryciini**

Tribe Blodeliini

Tribe Blodeliini was erected by Mensil (1960). According to Crosskey (1976), 18 genera have been reported from oriental region.

Key to Genera of Tribe Blodeliini

1. Eyes hairy. Scutellum with strong lateral setae 2
- Eyes bare. Scutellum with weak lateral setae sometimes absent 6

2. Propleuron bare. Facial ridges setose 3
– Propleuron hairy. Facial ridges bare...*Meigenia* R.D
3. Ocellar setae absent. Abdominal t_{1+2} excavate to its hind margin of corresponding segment ....*Compsilura* Bouche
– Ocellar setae present. Abdominal t_{1+2} not excavate to its hind margin of corresponding segment...4
4. Dorsocentral setae 3+3 5
– Dorsocentral setae 2+3 .... *Prosopofrontina* Town.
5. Vibrissae above level of epistomal margin. Parafacial with some hairs... *Biomeigenia* Mens.
– Vibrissae not above level of epistomal margin. Parafacial bare ....*Compsiluroides* Mens.
6. Wing cell r_5 open..7
– Wing cell r_5 closed..*Phytorophaga* Bezz.
7. Propleuron hairy...8
– Propleuron bare 9
8. Two presutural dorsocentral setae. Humeral callus with 3 setae standing in triangle. Last abdominal tergite normal with hairs fascicles...*Prodegeeria* B.B
– Three presutural dorsocentral setae. Humeral callus with 3 setae standing in straight line. Last abdominal tergite with sharp cone or very long tail without hairs fascicles .... ***Urodexia*** osten Sacken
9. Two post dorsocentral setae...10
– Three post dorsocentral setae 12
10. Midtibia with mid ventral setae. Fore tibia with two posterioventral setae .... *Degeeriopsis* Mens.
– Midtibia without mid ventral setae. Fore tibia with one posterioventral setae 11
11. In female, hindcoxa with some setae modified into short blunt black pegs .... *Medinodexia* Town.
– In female, hindcoxa with normal setae................................... *Medinomyia* Mens.
12. Two presutural dirsocentral setae. Abdominal t_{1+2} not excavate to its hind margin 13
– Three presutural dirsocentral setae. Abdominal t_{1+2} excavate to its hind margin 18
13. Gena very broad. Parafacials haired. Facial ridge setose............*Trichopaeia* B.B.
– Gena narrow. Parafacials bare. Facial ridge bare...14

14. Mid tibia with strong mid ventral setae. Facial ridge haired. Arista plumose or pubescent .... *Medina* R. D
- Mid tibia without mid ventral setae. Facial ridge bare. Arista bare..15
15. Abdominal t_3 and t_4 with discal setae .... *Trigonospila* Pok.
- Abdominal t_3 and t_4 without discal setae 16
16. One sternopleural setae. Presutural seta discaplaced inwards and standing near to presutural dorsocentral setae .... *Uroeuantha* Town.
- Two or three sternopleural setae. Presutural seta normal in position..............17
17. Arista with very long plumosity. Scutellum with strong lateral setae. Prosternum with few hairs............................... *Eophyllophila* Town.
- Arista with short plumosity. Scutellum with weak or without lateral setae. Prosternum with bare .... *Uromedina* Town.
18. Mid tibia with one mid dorsal setae. Fore tibia with 1 posterioventral setae 19
- Mid tibia with two mid dorsal setae. Fore tibia with 1 posterioventral setae 20
19. Apical scutellar setae absent .... *Gymnostylia* Wulp.
- Apical scutellar setae present... *Hemidegeeria* B.
20. Abdominal t_3-t_5 without discal setae... *Hygiella* Mens.
- Abdominal t_3-t_5 with discal setae.. *Euthelairosoma* B.

Genus – *Urodexia* Osten Sacken (1882)

From various parts of world more than 2 species of this genus have been reported. In India only 1 species has been reported. The genus is having following characters,

1. 3 presutural dorsocentral setae
2. Humeral callus with three setae standing in straight line
3. Frons are equally narrow
4. Last abdominal tergite of male shows sharp cone or very long tail.

Key to Species of the Genus *Urodexia*

1. Last visible abdominal tergite of male produced into cone .... ***uramyiodes***
- Last visible abdominal tergite of male produced into very long tail.. *penicillum*

Urodexia uramyioides **(Townsend, 1927)**

(Plate No.10, Figure 73 to 81)

Male (Figures 73-81)

Body 13.02 mm in length, 3.98mm in width; wing 10.12mm in length, 3.35mm in width; antenna 1.12mm long; halter 1.37mm length, 0.48mm width.

Head (Figure 75)

1.24mm in length, 3.04 mm in width; vertex narrow; inner vertical bristles with one reclinate setae; outer vertical bristle absent; ocelli present; ocellar bristles present; frontal vitta very narrow; frontal bristles medioclinate; fronto orbital plate with one additional row of medioclinate setae; reclinate upper orbital bristles with two strong setae; proclinate orbital bristles with two strong setae; face concave; lower facial margin not visible in lateral view; vibrissa strong; facial ridge concave; parafacial bare; gena narrow; genal dilation well developed; back of head slightly concave, with white hairs. **Eyes** 0.92mm bare, brownish. **Antenna** (Figure 76) scape 0.09mm in length, 0.11mm in width; pedicel 0.12mm in length, 0.15mm in width; first flagellomere 0.86mm in length, 0.14mm in width, concave. **Arista** inserted at base of first flagellomere; **Mouth** proboscis well developed; prementum with short labella; palpus well developed slightly swollen at apex.

Antennal Formula

S L/W=0.818, P L/W=0.80, F L/W=6.143, A=2.587

Thorax (Figures 77 and 78)

4.18mm in length, 3.20mm in width; humeral callus with 3 reclinate setae; proepisternum (propleuron) hairy; proepimeron with 3 setae dorsally projected; prosternum bare. **Scutum** 3.02mm in length, 3.17mm in width, 4 dark longitudinal strips; acrostichal bristles reclinate (presutural 3 + postsutural 3); dorsal central bristles reclinate setae (presutural 3+ postsutural 3), with few setulae; notopleuron with 2 reclinate setae; postalar callus with 2 reclinate setae. **Scutellum** 1.16mm in length, 2.13 mm in width; basal scutellar bristles slightly curved, short, lateral scutellar bristles as long as half of subapical scutellar bristles; subapical scutellar bristles divergent, longer; apical scutellar bristles divergent. **Subscutellum** convex; katepisternum with 2 strong setae (1 reclinate + 1 proclinate); anepimeron with 1 setae; katepimeron bare; meron with row of hairs; anatergite bare; anterior thoracic spiracle narrow and closed by fringes of hairs.

Wing (Figure 79)

10.12mm in length, 3.35mm in width, transparent; lower calypter well developed, inner margin less contiguous to lateral margin of scutellum; second costal section bare; base of costa with small setulae; costal bristles absent; fourth coastal section longer than 6th costal section; r_1 bare; cu a_1 bare; bend of with m obtuse angle; wing cell r_{4+5} open at wing margin; cross vein r-m present; cross vein d m-cu present; anal vein not reaching hind margin of wing. **Halter** reddish yellow, 1.37mm in length, 0.48mm in width.

Leg

Fore and mid femora with ventral spiny rows of bristles.**Forelegs** (Figure 80a) femora 4.14mm; tibia 3.42mm, tarsus 3.96mm, fore tibia preapical anteriodorsal setae present; fore tarsus laterally compressed, hard. **Midlegs** (Figure 80b) femora 4.03mm; tibia 3.38mm, tarsus 3.87mm mid tibia with anteriodorsal setae. **Hind legs** (Figure 80c) femora 4.08mm; tibia 3.45mm, tarsus 3.90mm, hind tibia with 1 anterodorsal setae; hind coxa with 1 small setae.

Abdomen (Figures 73 and 74)

7.78mm in length, 3.84 mm in width; last abdominal tergite produced into a sharp cone; mid dorsal depression reaches back to hind margin of that segment; discal bristles present, abdominal sternite overlapped by the ventral edges of tergite.

Genitalia (Figure 81)

Male terminalia retracted within sharp conus abdominal tergite 5th; both tergite 6th and 7th compressed laterally, overlapped by 5th tergite.

Colour

Yellowish brown: Palpus, thorax, abdomen.

Yellow: Fore tibia, mid tibia, hind tibia.

Silvery white: Head, face, parafacial, frontal vita, antenna.

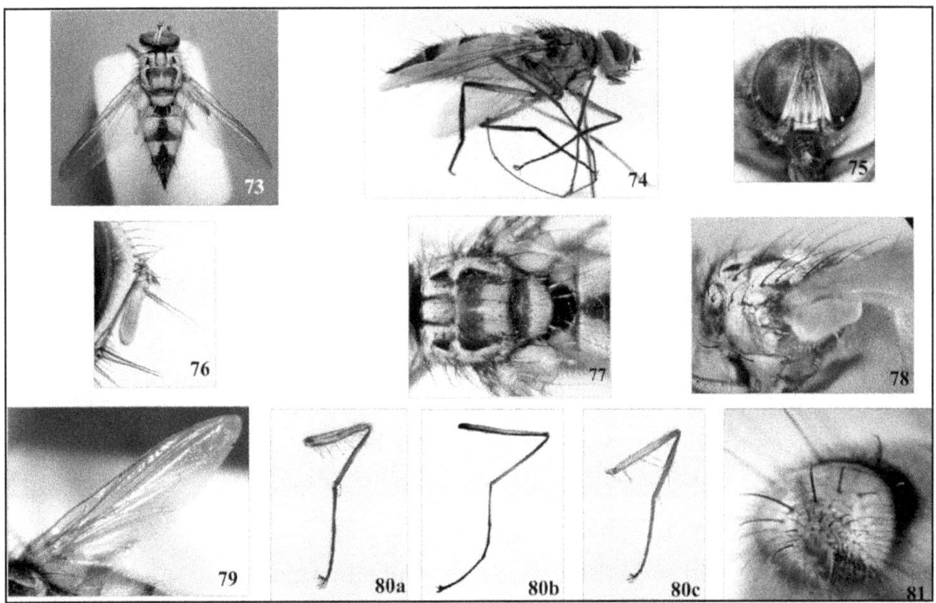

Plate 10. Figure 73: *Urodexia uramyioides* (Townsend) (Male) Dorsal view; Figure 74: Lateral View; Figure 75: Frontal View of Head; Figure 76: Antennae; Figure 77: Dorsal View of Thorax; Figure 78: Lateral View of Thorax; Figure 79: Fore Wing; Figure 80a: Fore Leg; Figure 80b: Mid Leg; Figure 80c: Hind Leg; Figure 81: Genitalia.

Reddish brown: Eyes.

Host- Unknown.

Host Plant- *Fragaria nubicola.*

Paratype- 4 Males, Sex ratio (M: F) 4:0, July 2011- February 2015; legs on slides, rest part pinned, preserved in insect box.

Distributional Record

India, Maharashtra, ♂ 1 Radhanagari 16- VII-2013; ♂1 Mahabaleshwar 11-XI-2013; ♂ 1 Gaganbawada 9-XI- 2014; ♂ 1 Amboli 20- VI-2015.

Remarks

According to key of Crosskey (1976) this species is *Urodexia uramyioides* (Townsend, 1927). Following characters are similar according to above key:

1. Last visible abdominal tergite of male produced into cone.
2. Hair fascicles are absent.

Additional characters refer to:

1. Pedicel with strong setae as 2x in diameter than vibrissae.
2. In some flies (paratypes) midlegs tarsus shows 5 tarsomeres.
3. **Antennal formula:** S L/W=0.818, P L/W=0.80, F L/W=6.143, A=2.587.

Tribe Exoristini

Tribe Exoristini was studied in oriental region by Chao (1964). According to Crosskey (1976), 10 genera have been reported from oriental region.

Key to Genera of Tribe Exoristini

1. Wing cell r_5 closed. 2+3 dorsocentral setae .... *Chaetoria* Beck.
– Wing cell r_5 open. 3+3 or 3+4 dorsocentral setae 2
2. Three post dorsocentral setae. Cross vein m-cu oblique .... *Stomatomyia* B.B.
– Four post dorsocentral setae. Cross vein m-cu not oblique 3
3. Upper occiput with row of well developed black setulae behind post postocular row .... *Chaetexorista* B.B
– Upper occiput without row of black setulae behind post postocular row......... 4
4. Facial ridges with strong downcurved setae. Eyes haired 5
– Facial ridges bare. Eyes haired or bare 9
5. Vibrissae inserted at the level of the epistomal margin 6
– Vibrissae inserted at the level above the epistomal margin 8
6. Abdominal tergite 3rd and 4th without median discal bristles .... ***Ctenophorinia*** Mens.
– Abdominal tergite 3rd and 4th with median discal bristles 7

7. Apical scutellar setae very strongly upwards .... *Phorinia* R.D
- Apical scutellar setae horizontal.. *Bessa* R.D
8. Wing with last section of cu_1 subequal in length to or shorter than m-cu.. *Austrophorocera* Town.
- Wing with last section of cu_1 longer than m-cu........................... *Chetogena* Rond.
9. Ocellar setae very strong. Bend of m with very long m_2 fold.. *Exorista* Meig.
- Ocellar setae small. Bend of m without long m_2 fold 10
10. Vibrissae inserted at the level of the epistomal margin .... *Phorcidella* Mens.
- Vibrissae inserted at the level above the epistomal margin...... *Eozenillia* Town.

Genus – *Ctenophorinia* Mensil (1963)

The genus *Ctenophorinia* was raised by Mensil (1963). According to this Tao *et al.* (2010) more than 2 species of genus have been reported. The genus is having following characters:

1. Eyes densely covered with long hairs.
2. Facial ridge with long, stout bristles.
3. Arista bare.
4. Palpi yellow.
5. Prosternum setose.
6. Mid coxa of female with strong spines.
7. Tergite 3rd and 4th without median discal bristles.

Key to Species of the Genus *Ctenophorinia*

1. Thorax with 3+3 arrangement of dorsocentral setae....*christianae* Zie. and shim
- Thorax with 3+4 arrangement of dorsocentral setae...2
2. Parafacials slightly wider than 1st flagellomere...............*frontalis* Zie. and shim.
- Parafacial as wide as first flagellomere..*dorsalis* sp. n

Ctenophorinia dorsalis sp.n.

(Plate 11, Figures 82 to 90)

Female (Figures 82-90)

Body 11.01 mm in length, 4.31 mm in width; wing 6.97mm in length, 3.94mm in width; antenna 1. 68 mm long; halter 1.17mm in length, 0.48mm in width.

Head (Figure 85)

1.64 mm in length,1.94mm in width; vertex very wide, Vertex is about 0.32 as wide as head in females; inner vertical bristles very strong, reclinate; pair of reclinate outer vertical bristle; ocelli present; ocellar bristles proclinate; frontal vita dark, well developed; frontal bristles medioclinate; fronto orbital plate golden colored with strong setae and row of medioclinate bristles; lateroclinate single setae very strong; proclinate orbital bristles with two pairs of setae; face concave; lower facial margin present not clearly visible in lateral view; vibrissa strong, crossed; facial ridge with strong row of setae at margin; parafacial entirely bare, silvery white, parafacial as wide as first flagellomere; gena narrow; genal dilation present; back of head with pale white hairs. **Eyes** 0.86mm in length, 1.56mm in width, hairy, dark brownish. **Antenna** (Figure 85) scape very short, 0. 11mm in length, 0.09mm in width, very short; pedicel 0.28mm in length, 0.15mm in width, short, with some hairs; first flagellomere 1.29mm in length, 0.14mm in width, slightly concave, 1st flagellomere about 4.3 times as long as pedicel. **Arista** long. **Mouth** proboscis well developed; prementum slightly elongated; palpus slightly swollen at apex.

Antennal Formula

S L/W=1.23, P L/W=1.8667, F L/W=9.215, A=4.104.

Thorax (Figures 86 and 87)

3.94mm in length, 3.11mm in width, shiny greenish pigments later on drying converts into golden yellow; black longitudinal strip of the thorax behind the suture narrows considerably towards the back and is extinguished long before postsutural

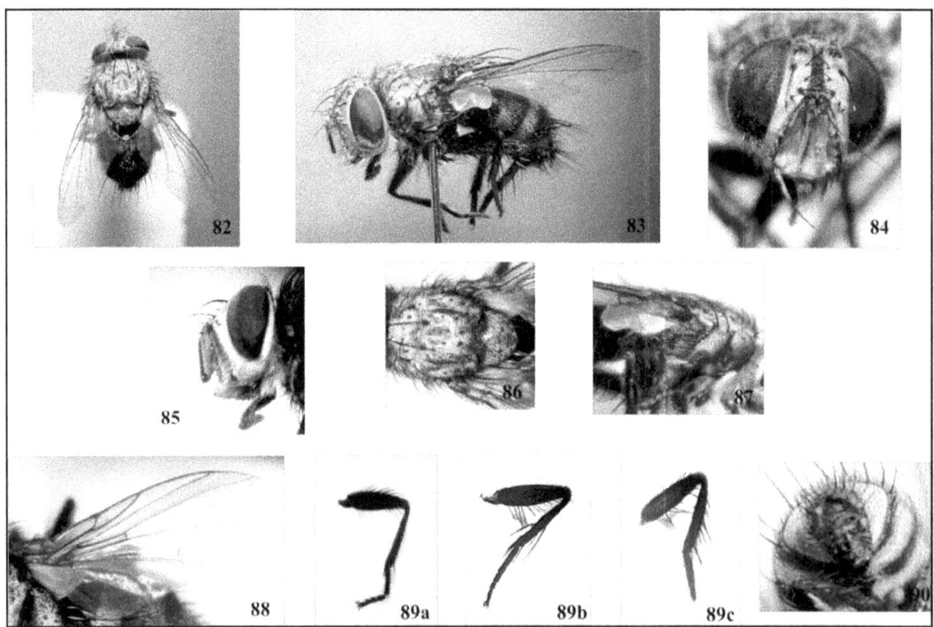

Plate 11. Figure 82: *Ctenophorinia dorsalis* sp.n. (Male) Dorsal view; Figure 83: Lateral View; Figure 84: Frontal View of Head; Figure 85: Antennae; Figure 86: Dorsal View of Thorax; Figure 87: Lateral View of Thorax; Figure 88: Fore Wing; Figure 89a: Fore Leg; Figure 89b: Mid Leg; Figure 89c: Hind Leg; Figure 90: Genitalia.

dorsocentral setae; humeral callus with three basal bristles standing in straight line; prosternum hairy; proepisternum bare; proepimeron with 2 setae. **Scutum** 2.81mm in length, 3.09mm in width; acrostichal bristles strong reclinate (presutural 3+ postsutural 3); dorsal central bristles strong, reclinate (presutural 3+ postsutural 4); notopleuron with 2 strong, reclinate setae; postalar callus with 2 reclinate setae, Black longitudinal stripe of thorax behind suture goes to last dorsocentral setae. **Scutellum** 1.13mm in length, 3.04mm in width; basal scutellar bristles present strong, slightly curved, lateral scutellar bristle present, long, strong, slightly curved; subapical scutellar bristles very long, strong; apical scutellar bristles parallel; subscutellum convex; katepisternum with 2 setae; anepimeron with 1 small setae; katepimeron bare; meron with row of setae; postmetacoxal area present; anatergite bare; anterior thoracic spiracle narrow and closed by fringes of hairs.

Wing (Figure 88)

6.97mm in length, 3.94mm in width, transparent, Basicosta dark brownish; lower calypter inner margin more contiguous to lateral margin of scutellum; second costal section bare; base of costa with small spine; costal bristles without strong costigeal bristles; fourth costal section longer than 6th; r_1 bare; cu a_1 bare; bend of m distinct; wing cell r_{4+5} open; cross vein r-m present; cross vein d m-cu present; anal vein not reaching towards hind margin of wing. **Halter** reddish yellow, 1.17mm length, 0.48mm in width.

Leg

Fore and mid femora with ventral row of spines ventrally.**Forelegs**(Figure 89a) femora 2.83mm; tibia 3.04mm and tarsus 2.11mm; fore tibia with 1 posterior setae; preapical anteriodorsal setae absent. **Midlegs** (Figure 89b) femora 2.87mm; tibia 3.09mm and tarsus 2.13mm; mid tibia with 2 anteriodorsal setae; 1 preapical anterior dorsal setae, 1 preapical posterioventral seta; mid coxa of female with 5 strong, slightly curved spines. **Hindlegs** (Figure 89c) femora 2.78mm; tibia 2.93mm and tarsus 2.05mm; hind tibia with 1 preapical anterioventral setae, 2 long preapical anteriodorsal seta; hind coxa bare.

Abdomen (Figures 82 and 83)

5.43mm in length, 3.08 mm in width, abdominal tergite 5th as long as 4th tergite; hairs from tergite 3rd and 4th dense; mid dorsal depression 1+2 syntergite, reaches towards 2 tergite; discal bristles present; abdominal sternites overlapped by ventrally.

Genitalia (Figure 90)

Terminalia retracted within abdominal tergite 5th; tergite 6th more reduced, joining segment 7th; tergite 7th and 8th fused and forms single segment (syntergite 7th + 8th); piercer present below syntergite 7th + 8th.

Colour

Black: Fore tibia, mid tibia, hind tibia, abdomen.

Golden Black: Thorax, parafrontal.

Brown: Antenna, palpus.

Dark brown: Frontal vita.

Silvery white: Head, face, parafacial.

Reddish brown: Eyes.

Host- Unknown.

Host Plant- *Symphyotrichum* sp., *Solidago canadensis.*

Holotype- Female, India, Maharashtra, Coll. 12-XI-2012, Chandgad, A.S. Desai; legs on slides, labeled as above.

Paratype- 4 Females, Sex ratio (M: F) 0:4, June- February (2011-2015).

Etymology- The species name is due to the most characters noted on dorsal side of species.

Distributional Record

India, Maharashtra, ♀ 1 Chandgad 12- IX-2012; ♀ 2 Panhala 19-XI- 2013; ♀ 1 Devgarh 30- VII-2015.

Remarks

According to key of Yao *et al.* (2010) this species runs close to *Ctenophorinia frontalis* (Ziegler, 1996) by having following characters:

1. Vertex is about 0.32, as wide as head in females.
2. 1st flagellomere about 4.3 times as long as pedicel.
3. Thorax with 3 presutural and 4 post sutural dorsocentral setae.

However, it differs from the above species by following distinct characters:

1. Color of parafrontal and thorax with shiny greenish pigments later on drying converts into golden yellow.
2. Parafacial as wide as first flagellomere.
3. Black longitudinal stripe of thorax behind suture goes to last dorsocentral setae.
4. Basicosta dark brownish.
5. Mid coxa of female with 5 strong, slightly curved spines.
6. Antennal formula: S L/W=1.23, P L/W=1.8667, F L/W=9.215, A=4.104.

Tribe Winthemiini

Tribe Winthemiini was studied by Baranov (1932) from oriental region. According to Crosskey (1976), 4 genera have been reported from oriental region.

Key to Genera of Tribe Winthemiini

3. Mid tibia with one mid dorsal setae. Two sternopleuron setae...........................2

– Mid tibia with several mid dorsal setae. Three sternopleuron setae 3

4. Parafacials bare .... *Nemorilla* Rond.
- Parafacials haired.. ***Winthemia*** R. D
8. Facial ridges strongly setulose. Two sternopleuron setae .... *Smidtiola* Mens.
- Facial ridges bare. Three sternopleuron setae .... *Timavia* R.D

Genus – *Winthemia* Robineau-Desvoidy (1830)

The genus ***Winthemia*** is raised by Robineau-Desvoidy (1830). According Rayner and Raper (2001) more than 9 species of this genus have been reported from Europe. In India only 1 species have been reported. The genus is having following characters,

1. Parafacial haired;
2. 2 sternopleural setae
3. Mid tibia with 1 anterodorsal setae.

Key to Species of the Genus *Winthemia*

1. Tergites 2 and 3 without marginal setae. 2 sternopleuron setae. Abdomen densely dusted. Mid tibia with 1 isolated anterodorsal setae 2
- Tergites 2 and 3 with marginal setae. 2 sternopleuron setae. Abdomen densely dusted. Mid tibia with 1 isolated anterodorsal setae 4
2. Tergite 5th with complete black in color with dense hairs. Frons as wide as 0.30-0.35x eye. 3rd antennal segment at most 2.5x to 3x as long as the 2nd antennal segment .... ***chandgadensis* sp.n**
- Tergite 5th with weakly black in color with few hairs. Frons as wide as 0.46-0.66x eye. 3rd antennal segment at most 2x to 2.6x as long as the 2nd antennal segment 3
3. Lowest row of posteriodorsal setae on hind tibia is not different from the posteriodorsal setae standing above it. Basal black color of tergite 5 not reaches the lateral edge ....*speciosa* Egg.
- Lowest row of posteriodorsal setae on hind tibia is longer and stronger posteriodorsal setae standing above it. Basal black color of tergite 5 reaches beyond the lateral edge....................................*venusta* Meig.
4. Three sternopleural setae. Tibia yellow. Tergites 2 and 3 without marginal bristles ....*bohemani* Zett.
- Two sternopleural setae. Tibia black. Tergites 2 and 3 with marginal bristles...5
5. Back of the head with black bristles behind post ocular hairs. Tergites 2 and 3 with irregular discal bristles 6
- Back of the head without black bristles behind post ocular hairs. Tergites 2 and 3 without discal bristles 7

6. Tergites dusted on their whole length..*variegata* Meig.
– Tergites dusted on their half of its length .... *jacentkovskyi* Mesn.
7. Marginal bristles of 4th tergite shorter than the segment, postalar callus yellow...8
– Marginal bristles of 4th tergite longer than the segment, postalar callus black 9
8. Frons 0.40- 0.49x as wide as one eye. Ocellar bristles longer than ocellar triangle...*cruentata* Rond.
– Frons 0.60- 0.65x as wide as one eye. Ocellar bristles shorter than ocellar triangle...*rufiventris* Macq.
9. Tergite 3 with 2-4 marginal setae ....*quadrpustulata* F.
– Tergite 3 with 4-6 marginal setae ....*erythrura* Meig.

Winthemia chandgadensis sp.n.

(Plate 12, Figures 91 to 99)

Male (Figures 91-99)

Body 12.19 mm long, 4.62 mm broad; wing 7.94mm in length, 3.68mm in width; antenna 1.21 mm long; halter 1.34mm in length, 0.48mm in width.

Head (Figures 93 and 94)

2.01 mm in length, 2.99mm in width; vertex wide; inner vertical bristles strong, long, reclinate; ocelli present; ocellar bristles proclinate; frontal vitta, present very well developed, with short hairs; frontal bristles medioclinate with small hairs; fronto orbital plate with row of setae; reclinate orbital setae absent; proclinate orbital setae absent; face concave; lower facial margin less visible slightly concave; vibrissa strong; facial ridge less visible, slightly concave; parafacial haired; gena narrow; genal dilation well developed; back of head with pale-white hairs. **Eyes** 1.27mm in length, 2.96mm in width; haired, reddish brown. **Antenna** (Figure 94) scape very short, 0. 07mm, in length, 0.11mm in width; pedicel 0.16mm in length, 0.14mm in width; first flagellomere 0.40mm in length, 0.13mm in width, concave. **Arista** bare, long, elongated. **Mouth** proboscis well developed; prementum short with small labella; palpus absent.

Antennal Formula

S L/W=0.636, P L/W=1.142, F L/W=3.076, A=1.618.

Thorax (Figures 95 and 96)

4.75mm in length, 3.52mm in width; black matt in color; with black longitudinal stripes; humeral callus with 5 reclinate setae; prosternum setulose; proepisternum bare; proepimeron with 2 setae. **Scutum** 3.44 mm in length, 3.49mm in width, with 4 black strips; acrostichal bristles (presutural 3+ postsutural 3); dorsal central bristles (presutural 3+ postsutural 4); notopleuron with 2 strong, reclinate setae;

intra alar setae 4 (1+3); postalar callus with 2 setae; Area of intraalar bristles from thorax with dust of color. **Scutellum** 1.31mm in length, 1.98mm in width; basal scutellar bristles slightly curved; lateral scutellar bristles weak; subapical scutellar bristles strong, divergent; apical scutellar bristles crossed; subscutellum convex; katepisternum with 2 setae; anepimeron single setae; katepimeron with small hairs; meron with row of setae; Golden bristles present behind regular row of black bristles on mesopleuron; anatergite with small hairs; anterior thoracic spiracle narrow and closed by fringes of hairs.

Wing (Figure 97)

7.94mm in length, 3.68mm in width; lower calypter well developed, hind margin less contiguous to margin of scutellum; second costal section bare; base of costa with small hairs; costal bristles not differentiated than costigeal setae; fourth costal section as long as 6th costal section; r_1 bare; cu a_1 bare; bend of m with distinct; wing cell r_{4+5} open; cross vein r-m present; cross vein d m-cu present; anal vein not reaches towards hind margin of wing. **Halter** reddish yellow, 1.34mm in length, 0.48mm in width.

Leg

Fore and mid femora with 2 rows of spines at ventrally; **Foreleg** (Figure 98a) femora 3.18mm; tibia 2.96mm and tarsus 3.06mm; with one small preapical

Plate 12. Figure 91: *Winthemia chandgadensis* sp.n. (Male) Dorsal view; Figure 92: Lateral View; Figure 93: Frontal View of Head; Figure 94: Antennae; Figure 95: Dorsal View of Thorax; Figure 96: Lateral View of Thorax; Figure 97: Fore Wing; Figure 98a: Fore Leg; Figure 98b: Mid Leg; Figure 98c: Hind Leg; Figure 99: Genitalia.

anteriodorsal setae, one small preapical posterioventral seta. **Mid legs** (Figure 98b) femora 3.10mm; tibia 2.88mm and tarsus 3.02mm with one anteriodorsal setae; one preapical anteriodorsal setae; one preapical posterioventral setae; one preapical posteriodorsal setae; one long preapical anterioventral, Posteriodorsal bristles shorter than bristles of the anteriodorsal comb. **Hind legs** (Figure 98c) femora 3.09mm; tibia 2.81mm and tarsus 3.03mm with comb like anteriodorsal setae; Lowest row of posteriodorsal bristles on hind tibia is not differentiated from the posteriodorsal bristles present above it; lowest posteriodorsal bristles is much longer and stronger than the bristles of anteriodorsal comb: one preapical posteriodorsal setae, one preapical posterioventral seta small; hind coxa with 1 hair.

Abdomen

5.43 mm in length, 3.12mm in width; abdominal tergite 4th as long as 5th. mid dorsal depression 1+2 syntergite reaches to 2 tergite margin; discal bristles absent; abdominal sternites overlapped by ventral edges of tergites; Tergite 4th with partially black color with silver- white dusting; black basal color of tergite 5th does not reach the lateral edge; tergite 5th with complete black color.

Genitalia (Figures 91 and 92)

Terminalia retracted within abdominal tergite 5th; tergite 6th more reduced to joining segment 7th. tergite 7th and 8th fused and forms single segment (syntergite 7th + 8th); cercus and surstylus present below syntergite 7th + 8th.

Colour

Black: Thorax, fore tibia, mid tibia, hind tibia.

Yellowish Brown: Abdomen, scutellum.

Brown: Frontal vita, antenna, palpus.

Silvery white: Head, face, parafacial.

Reddish brown: Eyes.

Host- Unknown.

Host Plant- *Nepeta indica.*

Holotype- Male, India, Maharashtra, Coll. 13-VIII- 2013, Panhala, A.S. Desai; antenna, wings, legs, tergite on slides, labeled as above.

Paratype- 4 Males, Sex ratio (M: F) 5:0, July, 2011- February, 2015; legs on slides, rest part pinned, preserved in insect box.

Etymology- The species name *chandgadensis* is described from the Chandgad region of Kolhapur, India where the type material has been collected.

Distributional Record

India, Maharashtra, ♂ 1 Panhala 13-VIII- 2013; ♂2 Chandgad 11-XI-2013; ♂ 1 Jath 10-XI- 2013; ♂ 1 Kagal 28- XI-2013.

Remarks

According to key of Crosskey, 1976 and Rayner and Raper, 2001, this species runs close to *Winthemia speciosa* (Egger, 1861) by having following characters,

1. Lowest row of posteriodorsal bristles on hind tibia is not differentiated from the posteriodorsal bristles present above it.
2. Posteriodorsal bristles shorter than bristles of the anteriodorsal comb.
3. Thorax looks black matt in color when viewed obliquely.
4. Area of intra alar bristles from thorax with dust of color.

However, it differs from the above species by following distinct characters,

1. Tergite 5th with complete black color.
2. Tergite 4th with partially black color with silver- white dusting.
3. Golden bristles present behind regular row of black bristles on mesopleuron.
4. Frons as wide as 0.30-0.35x of one eye.
5. **Antennal formula:** S L/W=0.636, P L/W=1.142, F L/W=3.076, A=1.618.

Tribe Carceliini

Tribe Carceliini was studied by Baranov (1934) from oriental region. According to Crosskey (1976), 7 genera have been reported from oriental region.

Key to Genera of Tribe Carceliini

1. Four post dorsocentral setae 2
- Three post dorsocentral setae 5

2. Eyes bare 3
- Eyes hairy .... ***Carcelia*** R.D

3. Four sternopleural setae 4
- Two to three sternopleural setae .... *Argyrophylax* B. B

4. Frontal setae all reclinate. Mid tibia with one anterodorsal setae .... *Thelyconychia* B.B
- Frontal setae normal. Mid tibia with two or more anterodorsal setae .... *Thecocarcelia* Town.

5. Eyes bare. Propleural setae absent. Hind tibia without posterodoesal setae .... *Hypersara* Vill.
- Eyes haired. Propleural setae present. Hind tibia with strong posterodoesal setae 6

6. Mid tibia with submedian vertical setae. Lateral scutellar setae weaker than basal setae .... *Argyrothelaira* Town.
- Mid tibia without submedian vertical setae. Lateral scutellar setae stronger than basal .... ***Carcelia*** R.D

Genus – *Carcelia* Robineau-Desvoidy (1830)

The genus is raised by Robineau-Desvoidy (1830). According to Rayner and Raper (2001) near about 14 species of this genus have been reported from europe. The genus shows following characters:

1. Four dorsocentral setae present.
2. Eyes haired.
3. Hind coxa bare or setulose.

Key to Species of the Genus *Carcelia*

1. Basicosta yellow. Middle tibia with one anterodorsal setae 2
– Basicosta black brown. Middle tibia with two to three anterodorsal setae 5
2. Humeral callus silvery white .... ***orientalis*** sp n.
– Humeral callus yellow or black 3
3. Humeral callus yellow. Dusting yellow grey to golden yellow .... *bombylans* R.D
– Humeral callus black. Dusting yellow grey to yellowish grey 4
4. Frons 0.30-0.40x as wide as one eye in male .... ***sataraensis*** sp n.
– Frons 0.42-0.72x as wide as one eye in male 5
5. Frons 0.42-0.50x as wide as one eye in male. Hairs of tergite 3 and 4 as long as 1/3 to 2/5 of the corresponding segment .... *rasa* Macq.
– Frons 0.64-0.72x as wide as one eye in male. Hairs of tergite 3 and 4 as long as 3/5 to 2/3 of the corresponding segment .... *puberula* Mesn.
6. Mid tibia yellow with one anterodorsal setae. Tergite 3rd and 4th haired and without discal setae 7
– Other combinations of features 8
7. Humeral callus black. Abdominal hairs rough. r_{4+5} with 2-3 bristles at base .... *atricosta* Hert.
– Humeral callus yellow. Abdominal hairs fine. r_{4+5} with 1 bristles at base .... *rasella* Bar.
8. Arista thickened for at least 2/5 of its length then tapering sharply. Abdomen haired 9
– Thickening of arista shorter in distance from base. Abdomen less hairs with few discal setae 10
9. Cheeks at mid point as wide as ½ -1/1 of 3rd antennal segment. Peristome as wide as 3rd antennal segment .... *iliaca* Ratz.
– Cheeks at mid point as wide as 1/5 -1/2 of 3rd antennal segment. Peristome narrower than 3rd antennal segment .... *gnava* Meig.

10. Apical scutellar setae shorter and weaker than lateral setae..............................11
- Apical scutellar setae stronger and longer than lateral setae............................12

11. 3rd antennal segment in males about 4x, in females about 3x as long as 2nd antennal segment..*tibialis* R.D
- 3rd antennal segment in males about 6-7x, in females about 4x as long as 2nd antennal segment...*falenaria* Rond.

12. Basicosta more or less brown yellow. Middle tibia completely yellow...*laxifrons* Vill.
- Basicosta black brown. Middle tibia blackened at some areas 13

13. Tibia black brown. Abdomen dusting faint.....................................*alpestris* Hert.
- Tibia yellow. Abdomen densely dusted..14

14. Tergites 3 and 4 with narrow black bands at posterior edge. Face in males as long as frons ....*kowarzi* Vill.
- Tergites 3 and 4 with very distinct black bands at posterior edge. Face is shorter than frons..15

15. Frons 0.50- 0.65x as wide as eye in males, 0.63- 0.79x in females ....*lucorum* Meig.
- Frons 0.44- 0.51x as wide as eye in males, 0.55- 0.72x in females ....*dubia* B.B.

Carcelia (Carcelia) orientalis sp.n.

(Plate 13, Figures 100 to 108)

Male (Figures 100-108)

Body 08.44 mm in length, 3.12 mm in width; wing 6.19mm in length, 2.59 mm in width; antenna 1.56 mm long, 0.23mm in width; halter 1.09mm in length, 0.38mm in width.

Head (Figures 102 and 103)

1.35mm in length, 2.57mm in width; vertex wide; inner vertical bristles single pair reclinate; outer vertical bristle present short reclinate; ocelli present; ocellar bristles strong, proclinate; frontal vita well developed; frontal bristles small, medioclinate; fronto orbital plate well developed, with 2 pairs of proclinate setae; two pairs of strong lateroclinate upper orbital bristles; proclinate orbital bristles two pairs; face concave, white; lower facial margin slightly visible in lateral view; vibrissa strong; facial ridge straight, not seen in lateral view; parafacial (cheeks) below frontal bristles bare, very narrow; gena not visible in the lateral view; peristome narrower; genal dilation not clearly visible; back of head with pale hairs. **Eyes** 0.78 mm in length, 2.78 mm in width, haired, reddish brown; **Antenna** (Figure 103) scape 0. 11mm in length, 0.12 mm in width, pedicel short, 0. 31mm in length, 0.15mm in width; first flagellomere 1.15mm in length, 0.14mm in width. **Arista** present near base of first flagellomere, cylindrically thickened; **Mouth** proboscis well developed; prementum with short labella; palpus well developed, slightly swollen apically.

Antennal Formula

S L/W=0.916, P L/W=2.06, F L/W=8.214, A=3.73.

Thorax (Figures 104 and 105)

3.51mm in length, 2.83mm in width; humeral callus with 3 setae, silvery white with black shade; proepisternum bare; proepimeron with 2 setae. **Scutum** 2.7mm in length, 2.79mm in width, with 4 strips; acrostichal bristles (presutural 2 + postsutural 3); dorsal central bristles (presutural 2 + postsutural 4); notopleuron with 2 reclinate setae; postalar callus with 2 setae. **Scutellum** 1.34 mm in length, 1.52mm in width; basal scutellar bristles very strong and slightly curved inwardly; lateral scutellar bristle small, weaker than basal setae; subapical scutellar very long, wide apart, strong, divergent; apical scutellar bristles very small, weak, crossed. **Subscutellum** convex; katepisternum with 2 sternopleuron setae; anepimeron with single setae; katepimeron bare; meron with row of setae; anatergite bare; anterior thoracic spiracle narrow closed with fringes hairs.

Wing (Figure 106)

6.19mm in length, 2.59 mm in width, transparent; lower calypter well developed its inner margin less contiguous to lateral margin of scutellum; second costal section

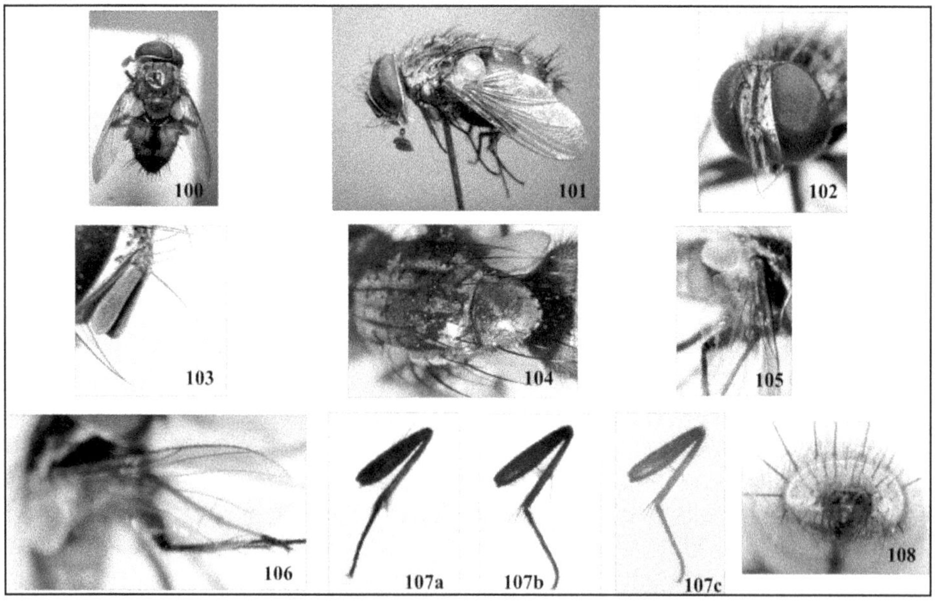

Plate 13. Figure 100: *Carcelia orientalis* sp.n. (Male) Dorsal view; Figure 101: Lateral View; Figure 102: Frontal View of Head; Figure 103: Antennae; Figure 104: Dorsal View of Thorax; Figure 105: Lateral View of Thorax; Figure 106: Fore Wing; Figure 107a: Fore Leg; Figure 107b: Mid Leg; Figure 107c: Hind Leg; Figure 108: Genitalia.

bare; base of costa with hairs, slightly lightened, Basicosta yellow; forth costal section longer than 6th section; r_1 bare; cu a_1 bare; bend of m distinct; wing cell r_{4+5} open, without hairs at base; cross vein r-m present; cross vein d m-cu present; anal vein present but not reaching towards hind margin of wing. **Halter** reddish yellow, 1.09mm length, 0.38mm width.

Leg

Fore and mid femora with row of spines ventrally; **Forelegs** (Figure 107a) femora 2.28mm; tibia 2.82mm and tarsus 2.17mm; preapical anteriodorsal setae present; fore tibia without bristle; **Mid legs** (Figure 107b) femora 2.23mm; tibia 2.76mm and tarsus 2.13mm; mid tibia with 1anterodorsal setae; preapical anteriodorsal setae; **Hind legs** (Figure 107c) femora 2.25mm; tibia 2.76mm and tarsus 2.14mm; hind tibia with pair of strong anteriodorsal setae; hind coxa setulose.

Abdomen (Figures 100 and 101)

3.85mm in length, 2.95mm in width, oval; 2nd abdominal tergite with short marginal setae; abdominal tergite 5th as long as 4th abdominal tergite; hairs of tergite 3rd and 4th covers 2/5 length of corresponding segment; marginal setae from 3rd tergite are equal in length of 4th tergite; mid dorsal depression reaching towards 1+ 2 syntergite; median discal bristles absent; abdominal sternites overlapped by ventral edges of tergites; 4th and 5th tergite dusted with yellowish grey.

Genitalia (Figure 108)

Terminalia retracted within abdominal tergite 5th; tergite 6th reduced attaches to terigite 7th and 8th; surstylus and cercus present below epandrium.

Colour

Black: Thorax, fore tibia, mid tibia, hind tibia.

Brown: Antenna.

Dark brown: Frontal vita.

Yellow: Abdomen, palpus.

Silvery white: Head, face, parafacial.

Reddish brown: Eyes.

Host- Unknown.

Host Plant- *Dacus carota, Tecoma castanifolia.*

Holotype- Male, India, Maharashtra, Coll. 03-VII- 2015, Gaganbawada, A.S. Desai; antenna, wings, legs, tergite on slides, labeled as above.

Paratype- 19 Males, Sex ratio (M: F) 19:0, July, 2011- February, 2015; legs on slides, rest part pinned, preserved in insect box.

Etymology- The species name is described from the oriental region *i.e.* Gaganbawada, India where the type material has been collected.

Distributional Record

India, Maharashtra, ♂ 4 Masayi Pathar (Panhala) 18- VIII-2012; ♂ 2 Panhala 15-VII- 2013; ♂ 2 Tikodi 06-XI- 2013; ♂ 1 Walawa 08-XI- 2013; ♂1 Sajjangad 21-XI-2013; ♂ 3 Mahabaleshwar 12-X- 2014; ♂ 6 Gaganbawada 03-VII- 2015.

Remarks

According to key of Crosskey (1976) and Rayner and Raper (2001) this species runs close to *Carcelia bombylans* (Robineau-Desvoidy, 1830) by having following characters:

1. Hairs of tergite 3rd and 4th covers 2/5 length of corresponding segment.
2. Basicosta yellow.
3. Middle tibia with one anteriodorsal setae.

However, it differs from the above species by following distinct characters,

1. Humeral callus silvery white.
2. 2nd abdominal tergite with short marginal setae.
3. Marginal setae from 3rd tergite are equal in length of 4th tergite.
4. 5th abdominal tergite is dorsoventrally compressed.
5. 4th and 5th abdominal tergite with dusting.
6. **Antennal formula:** S L/W=0.916, P L/W=2.06, F L/W=8.214, A=3.73

Carcelia (Carcelia) sataraensis sp.n.

(Plate 14, Figures 109 to 117)

Male (Figures 109-117)

Body 11.78 mm long, 1.21 mm broad, wing 8.03mm long, antenna 1.15 mm long; halter 1.49mm in length, 0.38mm in width.

Head (Figures 111 and 112)

1.82 mm in length, 3.39mm in width; vertex wide; inner vertical bristles paired, reclinate; outer vertical bristle absent; ocelli present; ocellar bristles absent; frontal vitta, dark well developed, narrow; frontal bristles medioclinate, descended more on parafacial; fronto orbital plate with very small hairs and row of medioclinate setae; lateroclinate with 2 setae; proclinate orbital bristles absent; face concave; frons 0.42-0.50x as wide as eye; lower facial margin less visible in lateral view; vibrissa present short; facial ridge slightly concave; parafacial bare, narrow; gena very narrow, bare; genal dilation present, scarcely developed; back of head concave, with few hairs pale white. **Eyes** 2.31mm in length, 1.05mm in width, brownish, hairy. **Antenna** (Figure 112) scape short, 0. 09mm in length, 0.11 mm in width pedicel 0.28mm in length, 0.14mm in width short; first flagellomere 0.78mm in length,0.14 mm in width slightly concave. **Arista** slightly long, slightly broader at base and elongated at apex. **Mouth** proboscis seldom, short; prementum short; palpus yellow, swollen at apex.

Antennal Formula

S L/W=0.818, P L/W=1.867, F L/W=5.572, A=2.752.

Thorax (Figures 113 and 114)

4.68mm in length; humeral callus with 3 reclinate setae, predominantly black; proepisternum bare; proepimeron with 2 setae; prosternum with some hairs. **Scutum** 3.42mm in length, 3.48mm in width, with 4 longitudinal strips; acrostichal bristles strong, reclinate (presutural 3+ postsutural 3); dorsal central bristles strong, reclinate (presutural 3+ postsutural 4); notopleuron with 2 reclinate setae; postalar callus with 2 setae. **Scutellum** 1.26mm in length, 3.38mm in width, yellow; basal scutellar bristles slightly curved, strong; lateral scutellar bristle divergent, long; subapical scutellar bristles slightly longer than all, divergent; apical scutellar bristles crossed, small; discal scutellar bristles present, small. **Subscutellum** convex; katepisternum with 2 setae; anepimeron 1 setae; katepimeron bare; meron with row of setae; anatergite bare; anterior thoracic spiracle narrow and closed by fringes of hairs.

Wing (Figure 115)

8.03mm in length, 3.13mm in width transparent; lower calypter well developed inner margin more contiguous to lateral margin of scutellum; second costal section

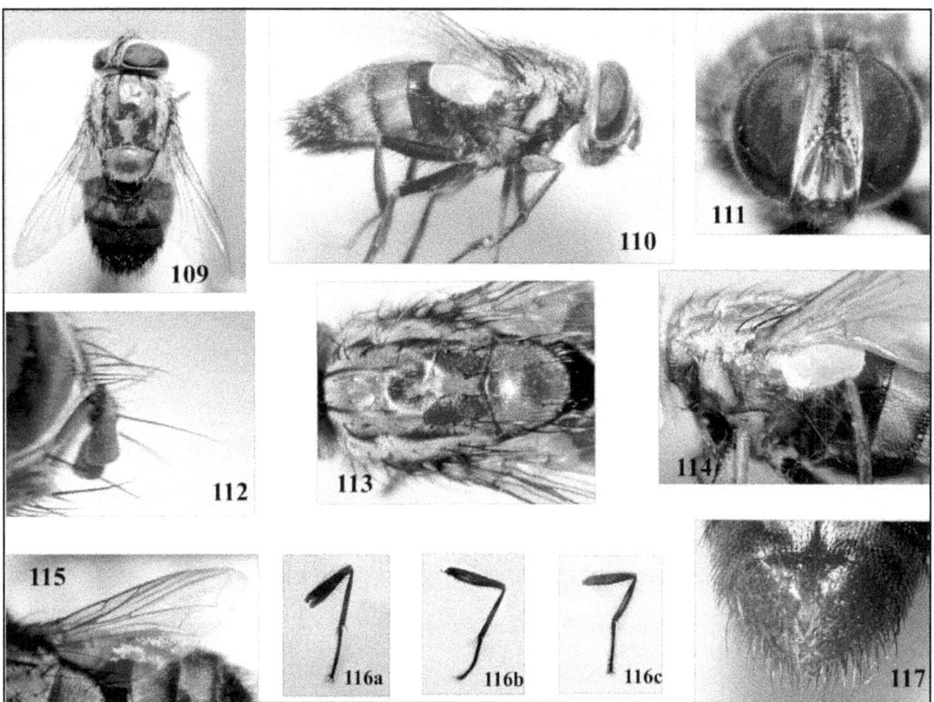

Plate 14. Figure 109: *Carcelia sataraensis* sp.n. (Male) Dorsal view; Figure 110: Lateral View; Figure 111: Frontal View of Head; Figure 112: Antennae; Figure 113: Dorsal View of Thorax; Figure 114: Lateral View of Thorax; Figure 115: Fore Wing; Figure 116a: Fore Leg; Figure 116b: Mid Leg; Figure 116c: Hind Leg; Figure 117: Genitalia.

bare; base of costa with few hairs, yellow; costal bristles not differentiated than costigeal setae; fourth costal section longer than 6th costal section; r_1 bare; cu a_1 bare; bend of m distinct; wing cell r_{4+5} open with few row of spines at base; cross vein r-m present; cross vein d m-cu present; anal vein not reaching towards hind margin of wing; **Halter** reddish yellow, 1.49mm in length, 0.38mm in width.

Leg

Fore and mid femora with row of spines ventrally. **Fore legs** (Figure 116a) femora 2.95mm; tibia 3.09mm and tarsus 3.19mm; fore tibia with 1 long preapical anteriodorsal setae, 1 anteriodorsal setae, 1 preapical posterioventral seta. **Mid legs** (Figure 116b)femora 3.02mm; tibia 3.15mm and tarsus 3.17mm; mid tibia with 1 anteriodorsal setae. **Hind legs** (Figure 116c) femora 2.91mm; tibia 3.04mm and tarsus 3.13mm; hind tibia with anteriodorsal present, anteriodorsal bristles present with few spines; hind coxa with hair.

Abdomen (Figures 113 and 114)

5.28mm in length, 3.53mm in width, grayish yellow; hairs of tergite 3rd and 4th are half the length of them; tergite 3rd with a pair of median marginal bristles; abdominal tergite 5th dark black with dense hairs, as long as 4th tergite; mid dorsal depression 1+2 syntergite, extending back to hind margin of that segment; discal bristles absent; abdominal sternites overlapped by ventral edges of tergites.

Genitalia (Figure 117)

Terminalia retracted within abdominal tergite 5th; tergite 6th more reduced, joining segment 7th; tergite 7th and 8th fused and forms single segment (syntergite 7th + 8th); cercus, surstylus present below syntergite 7th + 8th.

Colour

Black: Head, thorax, abdomen, fore tibia, mid tibia, hind tibia.

Brown: Antenna, Frontal vita.

Yellow: Palpus.

Silvery white: Face, parafacial.

Reddish brown: Eyes.

Host- Unknown **Host Plant-** *Dacus carota.*

Holotype- Male, India, Maharashtra, Coll. 12-X-2013, Wai, A.S. Desai; legs on slides labeled as above, rest part pinned, preserved in insect box.

Paratype- 4 Males, Sex ratio (M: F) 4:0, June 2011 to Feb. 2015; legs on slides, rest part pinned, preserved in insect box.

Etymology- The species name is due to the first record of the species from Satara district.

Distributional Record

India, Maharashtra, ♂ 2 Wai 12-X-2013; ♂ 1 Radhanagari 16-VIII- 2013; ♂ 1 Devgarh 04-XI- 2014.

Remarks

According to key of Rayner and Raper (2001) this species runs close to *Carcelia rasa* by having following characters:

1. Hairs of tergite 3rd and 4th are half the length of them.
2. Abdomen grayish yellow.

However, it differs from the above species by following distinct characters:

1. Frons 0.30-0.40x as wide as eye in male.
2. 5th abdominal tergites nearly black with dense hairs.
3. Apical scutellar bristles crossed.
4. Basal node of R_{4+5} with few hairs.
5. Tergite 3rd with a pair of median marginal bristles.
6. **Antennal formula:** S L/W=0.818, P L/W=1.867, F L/W=5.572, A=2.752.

Key to Sub-genera of Genus *Carcelia*

Genus *Carcelia* is divided into 6 subgenera given below:

1. *Carceliella* Baranov
2. *Senometopia* Macquart
3. *Catacarcelia* Townsend
4. *Carcelina* Mensil
5. *Euryclea* Robineau-Desvoidy
6. *Carcelia* Robineau-Desvoidy

Sub-genus *Senometopia*

The subgenus *Senometopia* was raised by Macquart (1834). According to Rayner and Raper (2001) near about 8 species of this sub-genus have been reported from Europe. The subgenus *Senometopia* shows following characters:

1. Mid tibia without ventral setae.
2. Humeral callus with three main setae standing in line.
3. Abdominal tergite without discal setae.
4. Hind coxa bare on posterodorsal surface.
5. Four dorsocentral setae.

Key to Species of the Sub-genus *Senometopia* Macquart (1834)

1. Three dorsocentral setae. Tibia black or brown.
   Dusting grey .... *Susurrans* Rond.

– Four dorsocentral setae. Tibia slightly yellow 2

2. Fore tibia with 2 posterior setae. Tergite 4 irregularly hairy and with discal setae. Dusting yellowish. Black longitudinal side strip of thorax behind the suture narrows not reaches to the last dc and does not become narrow .... *pollinosa* Mens.

– Fore tibia with 1 posterior setae. Tergite 4 irregularly hairy and with discal setae. Dusting yellowish. Black longitudinal side strip of thorax behind the suture broad reaches to the last dc and does not become narrow 3

3. Frons 0.66- 0.90x as wide as one eye in males, 0.78- 1.03x in females. 3rd tergite 2.3-2.9x as wide as their length. Hairs of tergites 3 and 4 less dense and irregular 4

– Frons 0.30- 0.76x as wide as one eye in males, 0.78- 0.89x in females. Hairs of tergites 3 and 4 few scattered 5

4. Dusting yellow. Black longitudinal side strip of thorax behind the suture narrows not reaches to the last dc and does not become narrow .... *intermedia* Hert.

– Dusting grey. Black longitudinal side strip of thorax behind the suture narrows not reaches to the last dc and does not become narrow .... *confundens* Rond.

5. Frons 0.25- 0.45x as wide as one eye in males .... ***kolhapurensis*** sp. n

– Frons 0.51- 0.70x as wide as one eye in males .... *separata* Rond.

***Carcelia* (*Senometopia*) *kolhapurensis* sp.n.**

(Plate 15, Figures 118 to 126)

Male (Figures 118-126)

Body 9.93mm long, 3.20mm broad; wing 7.30mm in length, 3.06mm in width; antenna 1. 36 mm in length; halter 1.14mm in length, 0.41mm in width.

Head (Figures 120 and 121)

1.92mm in length, 3.42mm in width; vertex wide; inner vertical bristles strong, reclinate; pair of reclinate outer vertical bristle; ocelli present; ocellar bristles proclinate; frontal vita dark, well developed; frontal bristles medioclinate; fronto orbital plate with medioclinate bristles, yellowish dusting along with the edge of frontal vita; lateroclinate single setae very strong; proclinate orbital bristles with two pairs of setae; face concave; lower facial margin present not clearly visible in lateral view; vibrissa strong, crossed; facial ridge with strong row of setae at margin; parafacial with 3 to 4 bristles found below the base of antennae, silvery white; gena narrow; genal dilation weak; back of head with pale white hairs. **Eyes** 1.53mm in length, 2.43mm in width, with hairs, dark brownish. **Antenna** (Figure 121) scape 0.12mm in length,0.10mm in width, pedicel 0.23mm in length, 0.13mm in width; first flagellomere 1.01mm in length,0.15 mm in width, slightly concave. **Arista** long, bare. **Mouth** proboscis well developed; prementum slightly elongated; palpus slightly swollen at apex.

Antennal Formula

S L/W=1.2, P L/W=1.769, F L/W=6.733, A= 3.234.

Thorax (Figures 122 and 123)

3.87mm in length, 2.69mm in width; humeral callus with 3 basal bristles standing in straight line; prosternum hairy; proepisternum bare; proepimeron with 2 setae. **Scutum** 2.75mm in length, 2.65mm in width; Black longitudinal side strip of thorax behind the suture goes to the last dc and does not become narrow; acrostichal bristles reclinate (presutural 3+ postsutural 3); dorsal central bristles strong, reclinate (presutural 3+ postsutural 4); notopleuron with 2 strong, reclinate setae; postalar callus with 2 reclinate setae. **Scutellum** 1.11mm in length, 1.45mm in width, grayish; basal scutellar bristles present strong, slightly curved, lateral scutellar bristle present, long, strong, slightly curved; subapical scutellar bristles very long, strong; apical scutellar bristles parallel; subscutellum convex; katepisternum with 2 setae; mesopleuron with single setae present near edge of humeral callus; anepimeron with 1 small setae; katepimeron bare; meron with

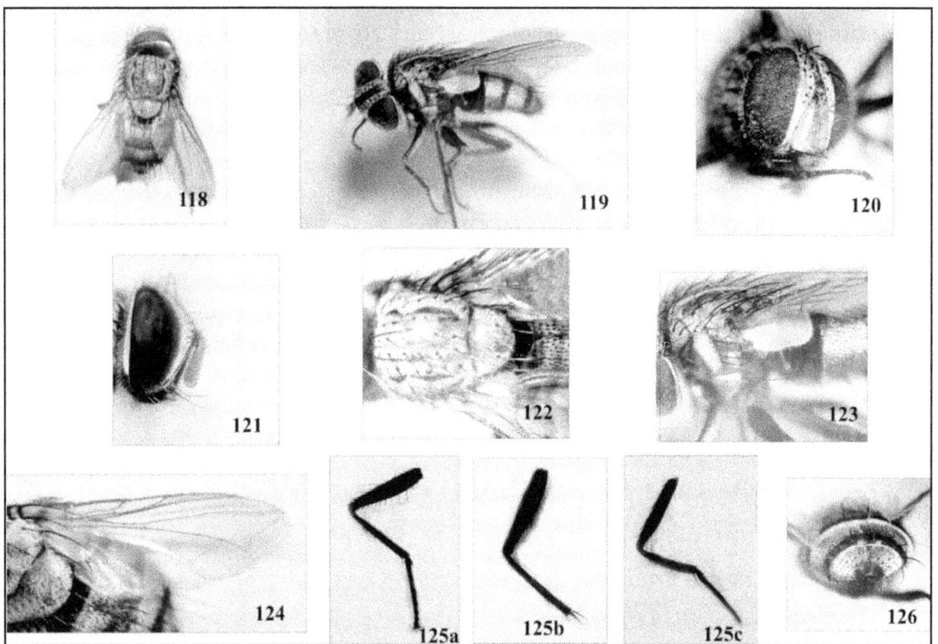

Plate 15. Figure 118: *Carcelia* (*Senometopia*) *kolhapurensis* sp.n. Dorsal view; Figure 119: Lateral View; Figure 120: Frontal View of Head; Figure 121: Antennae; Figure 122: Dorsal View of Thorax; Figure 123: Lateral View of Thorax; Figure 124: Fore Wing; Figure 125a: Fore Leg; Figure 125b: Mid Leg; Figure 125c: Hind Leg; Figure 126: Genitalia.

row of setae; postmetacoxal area present; anatergite bare; anterior thoracic spiracle narrow and closed by fringes of hairs.

Wing (Figure 124)

7.30mm in length, 3.06mm in width, transparent; lower calypter inner margin more contiguous to lateral margin of scutellum; second costal section bare; base of costa with small spine; costal bristles without strong costigeal bristles; fourth costal section longer than 6th; r_1 bare; cu a_1 bare; bend of m distinct; wing cell r_{4+5} open; cross vein r-m present; cross vein d m-cu present; anal vein not reaching towards hind margin of wing. **Halter** reddish yellow, 1.14mm in length, 0.41mm in width.

Leg

Fore and mid femora with ventral row of spines ventrally; **Forelegs** (Figure 125a) foretibia 2.48mm; tibia 2.97mm and tarsus 2.42mm; 1 posterior setae; preapical anteriodorsal setae absent. **Midlegs** (Figure 125b) midtibia 2.44mm; Tibia 3.02mm and tarsus 2.33mm; with 2 anteriodorsal setae; 1 preapical anterior dorsal setae, 1 preapical posterioventral seta; **Hindlegs** (Figure 125c) hind tibia 2.41mm; tibia 3.00mm and tarsus 2.27mm;1 preapical anterioventral setae, 2 long preapical anteriodorsal seta; hind coxa posteriodorsally bare.

Abdomen (Figures 118 and 119)

4.14mm in length, 3.47mm in width; dusting grey color on each tergite; abdominal tergite 5th as long as 4th tergite; hairs from tergite 3rd and 4th less dense; mid dorsal depression 1+2 syntergite, reaches towards 2 tergite; discal bristles present; abdominal sternites overlapped by ventrally.

Genitalia (Figure 126)

Terminalia retracted within abdominal 5th tergite; tergite 6th less reduced joining segment 7th; surstylus and cercus present.

Colour

Black- Fore tibia, mid tibia, hind tibia, abdomen.

Golden Black: Thorax, parafrontal.

Brown: Antenna, palpus.

Dark brown: Frontal vita.

Silvery white: Head, face, parafacial.

Reddish brown: Eyes.

Host- *Helicoverpa armigera*

Host Plant- *Ruellia brittoniana.*

Holotype- Male, India, Maharashtra, Coll. 12- IX-2012, Chandgad, A.S. Desai; legs on slides, labeled as above.

Paratype- 5 Males, Sex ratio (M: F) 5:0, July 2011- February 2015; legs on slides, rest part pinned, preserved in insect box.

Etymology- The species name is described from the Kolhapur, India where the type material has been collected.

Distributional Record

India, Maharashtra, ♀ 1 Chandgad 12- IX-2012; ♀ 2 Panhala 19-XI- 2013; ♀1 Mahabaleshwar 01-X-2014; ♀ 1 Devgarh 30- VII-2015.

Remarks

According to key of Crosskey (1976) and Rayner and Raper (2001), this species runs close to *Senometopia separata* (Rondani, 1859) by having following characters:

1. Dusting of grey color on each tergite.
2. Black longitudinal side strip of thorax behind the suture goes to the last dc and does not become narrow.

However, it differs from the above species by following distinct characters:

1. Hairs from abdominal tergite 3rd and 4th with less dense.
2. Mesopleuron with single setae present near edge of humeral callus.
3. Parafacials with 3 to 4 bristles found below the base of antennae.
4. Scutellum grayish in color.
5. Parafrontal with yellowish dusting along with the edge of frontal vita.
6. **Antennal formula:** S L/W=1.2, P L/W=1.769, F L/W=6.733, A= 3.234.

Tribe Sturmiini

Tribe Sturmiini was studied by Mesnil (1949-1952) from oriental region. According to Crosskey (1976), 23 genera have been reported from oriental region.

Key to Genera of Tribe Sturmiini

1. Head with 2 or three pairs of reclinate orbital setae 2
– Head with 1 pairs of reclinate orbital setae 19
2. Eyes densly haired 3
– Eyes bare 8
3. Parafacials haired .... *Sturmiopsis* Town.
– Parafacials bare 4
4. Basal node of r_{4+5} with at least 2 -3 small setulae. Facial ridge normal 5
– Basal node of r_{4+5} with long and strong setula.
Facial ridge normal .... *Isochaetina* Mesn.
5. Facial ridges with very strong downcurved setae .... *Takanomyia* Mesn.
– Facial ridges bare 6
6. Two to three sternopleural setae 7
– Four sternopleural setae .... *Paradrino* Mesn.

7. Mid tibia with one or two anterodorsal setae. Prescutum and scutum bright yellow pollinose .... *Calozenillia* Town.
- Mid tibia with three or more anterodorsal setae. Prescutum and scutum bright without yellow pollinose .... *Euhygia* Mesn.
8. Mid tibia with 2 or more anterodorsal setae...9
- Mid tibia with one anterodorsal setae 14
9. Two to three sternopleural setae 10
- Four (2+2) sternopleural setae 13
10. Prosternum bare. Facial ridge with down curved setae ....*Blepharella* Macq.
- Prosternum haired. Facial ridge bare or with weak setae 11
11. Antennal axis about level with the eye middle. Parafacials narrowing towards lower end 12
- Antennal axis above level with the eye middle. Parafacials not narrowing towards lower end ....*Pexopsis* B.B
12. Ocellar setae absent. Eyes more or less bare.................................. *Pujolina* Mesn.
- Ocellar setae strong. Eyes with small hairs.................................... *Euhygia* Mesn.
13. Ocellar setae strong. Basal node of r_{4+5} with one setula ....*Zygobothria* Mik.
- Ocellar setae hairs like, absent. Basal node of r_{4+5} with two setula .... *Weingaertneriella* Bor.
14. Three sternopleural setae. Ocellar setae strong...15
- Four sternopleural setae. Ocellar setae absent or weak 16
15. Second supra alar setae reduced. Apical scutellar setae horizontal. Palpi blackish .... *Cadurcia* Vill.
- Second supra alar setae normal. Apical scutellar setae directed half upwards. Palpi blackish ....*Thelairodrino* Mesn.
16. Basal node of r_{4+5} with 3 or more setulae. Ocellar setae absent or minute ....*Isosturmia* Town.
- Basal node of r_{4+5} with one strong setulae. Ocellar setae varied 17
17. Ocellar setae very strong. Scutellum with one pair of very strong lateral setae. Parafacial bare...*Zygobothria* Mik.
- Ocellar setae weak or absent. Scutellum with two pair of very strong lateral setae. Parafacial bare or hairy...18
18. Parafacial bare. Ocellar setae absent ....*Drino* R.D
- Parafacial haired. Ocellar setae present in some flies............... *Palexorista* Town.
19. Eyes densely haired...20
- Eyes bare 23

20. Parafacials bare. Scutellum with lateral setae..21
– Parafacials haired. Scutellum without lateral setae........................*Tritaxys* Macq.
21. Scutellum with two or three pairs of discal setae ....*Trixomorpha* B.B
– Scutellum with one pairs of discal setae 22
22. Facial ridges with very strong down curved setae................................ *Pales* R.D.
– Facial ridges bare ....*Sisyropa* B.B
23. Subapical scutellar setae widely separated from each other. One pair of lateral scutellar setae. Four sternopleural setae .... *Sturmia* R.D
– Subapical scutellar setae not widely separated from each other. One pair of lateral scutellar setae. Two to four sternopleural setae..................24
24. Facial ridges weakly setose on the lower half. Humeral callus with three setae. 2 sternopleural setae. Ocellar setae absent.......*Parapales* mesn.
– Facial ridges bare except few setulae. Humeral callus with four setae. 3 to 4 sternopleural setae. Ocellar setae absent or absent................25
25. Two to three sternopleural setae. Parafacial hairy. Large forms..***Blepharipa*** Rond.
– Two to three sternopleural setae. Parafacial hairy. Large forms......*Sisyropa* B.B

Genus – *Blepharipa* Rondani (1856)

The Genus *Blepharipa* was erected by Rondani (1856). From various parts of world more than 17 species of this genus have been reported. In India only 4 species have been reported. The genus shows following characters:

1. Subapical scutellar setae not widely separated.
2. 2-3 sternopleural setae present.
3. Parafacial with some hairs.
4. Length always between 10-20mm.

Key to Species of the Genus *Blepharipa* Rondani (1856)

1. Tergite 2 and 3 with marginal setae. Dusting of tergite covers complete segment length ....*pratensis* Meig.
– Tergite 2 and 3 without marginal setae. Dusting of tergite not covers complete segment length..***schineri*** Mesn.

Blepharipa schineri (Mesnil, 1939)

(Plate 16, Figures 127 to 135)

Male (Figures 127-135)

Body 13.36 mm in length, 5.25 mm in width; wing 11.9 mm in length, 4.72mm in width; antenna 1.16 mm in length, 0.23mm in width; Halter 1.31mm in length, 0.61mm in width.

Head (Figures 129 and 130)

4.30mm in length, 2.61mm in width; vertex slightly wide; inner vertical bristles single; outer vertical bristle one on lateral side; ocelli present; ocellar bristles small with proclinate manner; frontal vita dark brown, wide; frontal bristles 10 pairs descending on base of antenna; fronto orbital plate with scattered hairs; lateroclinate upper orbital bristles with single pair; proclinate orbital bristles three pairs; face silvery white, concave; lower facial margin not visible in lateral view; vibrissa well developed, strong, arising from lower facial margin; facial ridge slightly convex; parafacial entirely with fine hairs; gena wider than profrons; genal dilation well developed, fine and short hairs present at the base of vibrissae; back of head convex, with pale-yellow hairs. **Eyes-** 3.41mm in length, 1.32mm in width, bare. **Antenna** (Figure 130)- scape short, 0.11mm in length, 0.15mm in width; pedicel short, 0.28mm in length, 0.16mm in width; first flagellomere 0.77mm in length, 0.12 in width, concave. **Arista** inserted near the base of first flagellomere; **Mouth-** proboscis small but well developed; prementum short with swollen labella; palpus well developed.

Antennal Formula

S L/W=0.733, C L/W=1.75, F L/W=6.416, A=2.981.

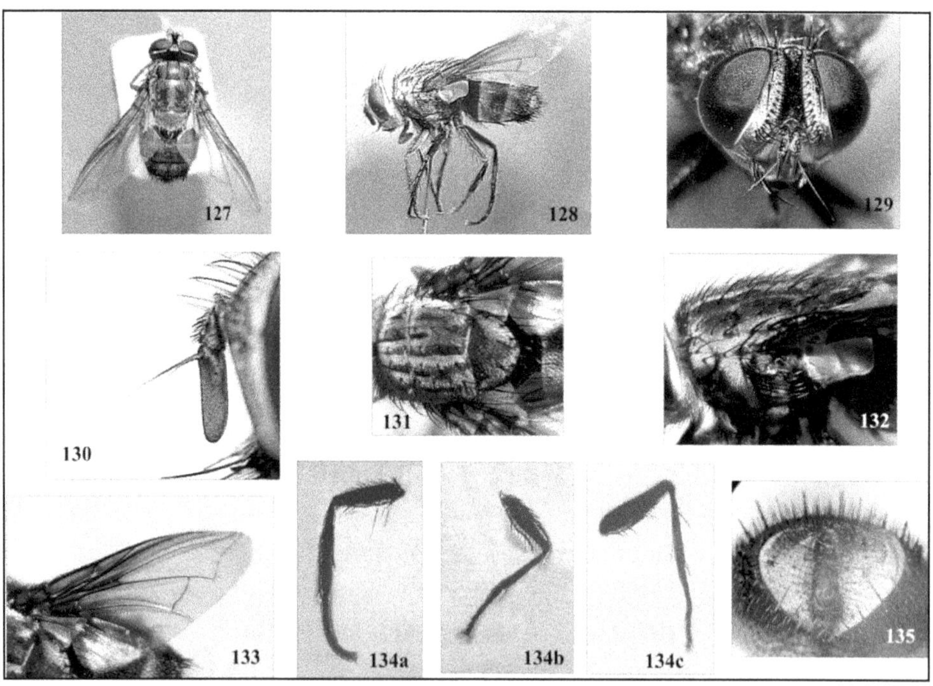

Plate 16. Figure 127: *Blepharipa schineri* Mesnil (Male) Dorsal view; Figure 128: Lateral View; Figure 129: Frontal View of Head; Figure 130: Antennae; Figure 131: Dorsal View of Thorax; Figure 132: Lateral View of Thorax; Figure 133: Fore Wing; Figure 134a: Fore Leg; Figure 134b: Mid Leg; Figure 134c: Hind Leg; Figure 135: Genitalia.

Thorax (Figures 131 and 132)

6.19mm in length, 4.96mm in width, black with thin silvery pollinosity most evident on pleural regions; humeral callus with 4 setae; proepisternum with small hairs; proepimeron with 2 prostagmitic setae and 2 propleural setae; prosternum with some hairs. **Scutum** 4.08mm in length, 4.91mm width; with 2 dark broad strips; acrostichal bristles present (presutural 3+ postsutural 3); dorsal central bristles present (presutural 3+ postsutural 4); notopleuron with 2 setae; postalar callus with 3 setae. **Scutellum** 2.11 mm in length, 3.24mm in width; basal scutellar bristles strong, shorter than lateral scutellar strong and longer than lateral scutellar bristle; subapical scutellar bristles more longer and stronger than other, apical scutellar bristles strong, short, crossed. **Subscutellum** convex, bare; katepisternum with 3 strong setae; anepimeron with some strong hairs at the base of wing; katepimeron bare; meron small setulae; postmetacoxal area present; anatergite bare; anterior thoracic spiracle narrow closed with small hairs.

Wing (Figure 133)

11.9 mm in length, 4.72mm in width; lower calypter well developed its inner margin contiguous to lateral margin of scutellum; second costal section with fine hairs; base of costa without costigial bristles; r_1 bare; cu a_1 bare; bend of m distinct; wing cell r_{4+5} open; cross vein r-m present; cross vein d m-cu present; anal vein present, but not reaching towards hind margin of wing. **Halter** reddish yellow, 1.31mm in length, 0.61mm in width.

Leg

Forelegs (Figure 134a) femora 3.64mm; tibia 3.23mm and tarsus 2.55mm; femora with two rows of spines ventrally; fore tibia with one preapical anteriodorsal setae longer than preapical dorsal setae. **Midlegs** (Figure 134b) femora 3.36mm; tibia 3.18mm and tarsus 2.49mm; mid tibia with one anterior dorsal setae. **Hindlegs** (Figure 134c) femora 3.62mm; tibia 3.20mm and tarsus 2.51mm; hind tibia with single row posterior ventral setae; one preapical anteriodorsal setae, 2 preapical posterioventral seta; hind coxa with 2 long setae and 2 small setae.

Abdomen (Figures 127 and 128)

5.39mm in length, 5.78mm in width, oval, shiny; abdominal tergite 2nd and 3rd without marginal bristles; abdominal tergite 4th and 5th are equal in length; sides of abdominal tergite 4th with long hairs; mid dorsal depression extending back to hind margin of 2 segments with integral sutures; discal bristles present including 1+2 syntergites; abdominal sternites overlapped by ventral edges of tergites.

Genitalia (Figure 135)

Terminalia retracted within abdominal tergite 5th; tergite 6th more reduced, joining segment 7th; tergite 7th and 8th fused and forms single segment (syntergite 7th + 8th); cercus and surstylus are present below tergite 7th and 8th.

Host- Unknown.

Host Plant- *Tecoma castanifolia, Aster amellus.*

Paratype- Sex ratio (M: F) 4:0, July 2011- February 2015; legs on slides, rest part pinned, preserved in insect box.

Distributional Record

India, Maharashtra, ♂ 1 Gargoti 20- VIII-2012; ♂1 Miraj 13-VII-2013; ♂ 1 Patan 01-X- 2014; ♂ 1 Dahiwadi 09-XI- 2014.

Remarks

According to key of Crosskey (1976) and Rayner and Raper (2001) this species is *Blepharipa schineri* (Mensil, 1939). Following characters are similar according to above key:

1. Tergite 2 and 3 without marginal setae.
2. Dusting of tergite not covers (2/3- 4/5) complete segment length.

Additional characters are as follows:

1. Tergite 4th always 2.5 to 3x of 5th tergite.
2. Frons 2.5 to 3x as wide as one eye.
3. **Antennal formula:** S L/W=0.733, C L/W=1.75, F L/W=6.416, A=2.981.

Tribe Eryciini

Tribe Eryciini was studied by van Embden (1954) from oriental region. According to Crosskey (1976), near about 38 genera have been reported from oriental region.

Key to Genera of Tribe Eryciini

1. Three post dorsocentral setae 2
- Four post dorsocentral setae 6
2. Eyes haired. Frontal setae not reclinate .... *Rhinomyodes* Town.
- Eyes bare. Frontal setae completely reclinate 3
3. Parafacials haired. Ocellar setae strong. Abdomen with thick yellow pollinosity .... *Buquetia* R.D
- Parafacials bare. Ocellar setae weak. Abdomen with thick greyish pollinosity 4
4. Wing cell with r_5 closed. Mid tibia with two or more anteriodorsal setae .... *Simosa* Aldr.
- Wing cell with r_5 open narrowly. Mid tibia with one anteriodorsal setae 5
5. Abdomen with median discal setae on t_3 and t_4. Basal node of vein r_{4+5} with one setula .... *Elodimyia* Mes.
- Abdomen without median discal setae on t_3 and t_4. Basal node of vein r_{4+5} with row of 3 to 7 setulae .... ***Aneogmena*** B.B.

6. Wings with vein cu_1 setose basally, vein r_{4+5} setulose on most of its length and vein r_1 setulose 7
- Wings with vein cu_1 bare, vein r_{4+5} setulose on basal node and vein r_1 bare 8
7. Mesonotum with 3+3 acrostichal setae.................... *Prosopodopsis* Town.
- Mesonotum with reduced 1+1 acrostichal setae .... *Suensonomyia* Mesn.
8. Wing cell r_5 closed. 2 presternal dorsocentral setae....................9
- Wing cell r_5 open. 3 presternal dorsocentral setae 10
9. Head strongly triangular. One pair of reclinate orbital setae. Apical scutellar setae horizontal....................*Diatraeophaga* Town.
- Head strongly triangular. One pair of reclinate orbital setae. Apical scutellar setae horizontal....................*Cestonia* Rond.
10. Mid tibia with mid ventral seta. Barette bare or with small hairs. Pre alar setae stronger than the first post intra alar setae....................11
- Mid tibia without mid ventral seta. Barette strong hairs. Pre alar setae weaker than the first post intra alar setae.......*Bactromyiella* Mesn.
11. Eyes haired....................21
- Eyes bare 12
12. Four sternopleural setae (2+2)....................13
- Two to three sternopleural setae 14
13. Facial ridge bare. Apical scutellar setae horizontal. Abdominal t_3 and t_4 without discal setae .... *Aplomya* R.D
- Facial ridge setulose. Apical scutellar setae directed partially upwards. Abdominal t_3 and t_4 without discal setae ....*Pseudoperichaeta* B.B.
14. Parafacial hairy on upper half. Ocellar setae almost absent. Arista thickened on nearly all its length.................... *Botriopsis* Town.
- Parafacial bare. Ocellar setae present. Arista thickened not more than half its length....................15
15. Humeral callus with 3 setae standing in triangle 16
- Humeral callus with 3 setae standing in a straight line 19
16. Epistome easily visible. Palpi long, yellow .... *Phinaplomyia* Mens.
- Epistome normal, mostly invisible. Palpi normal, blackish 17
17. Abdominal t_3 with four median marginal setae. 3 sternopleuron setae .... *Phebellia* RD
- Abdominal t_3 with two median marginal setae. 2 sternopleuron setae. ....*Exorista seniorwhitei* Barn.

(Generic position of this species problematic)

18. Facial ridges more or less setulose. Abominal t $_3$ and t_4 with erect median discal setae. Antennal axis above the level of eye middle....................19
- Facial ridges bare. Abominal t_3 and t_4 without erect median discal setae. Antennal axis at the level of eye middle .... *Neasomyia* Mesn.
19. Mid tibia with several strong anterodorsal setae. Apical scutellar setae strong. Palpi black. Hairing of eyes long and dense .... *Phryxe* R.D.
- Mid tibia with one anterodorsal setae. Apical scutellar setae weak. Palpi yellow. Hairing of eyes short .... *Zenillia* R.D
20. Parafacials bare 21
- Parafacials hairy..*Pseudalsomyia* Mesn.
21. Vein r_1 haired 22
- Vein r_1 bare 24
22. Legs reddish yellow. Basicosta yellow. Humeral callus with row of 3 setae...23
- Legs reddish black brown. Basicosta brown. Humeral callus with row of 2 setae...*Eurysthaea* R.D
23. Facial ridge bare .... *Diglossocera* Wul.
- Facial ridge more or less setulose...*Hapalioloemus* Bar.
24. Scutellum without lateral setae. Head with one pair of reclinate orbital setae ....*Dolichocolon* B.B.
- Scutellum with lateral setae. Head with two to three pair of reclinate orbital setae...25
25. Suapical setae reduced and apical setae enlarged ....*Atractocerops* Town.
- Suapical setae normal and apical setae weak 26
26. Facial ridges with strong downcurved setae. Mid tibia with several strong anterodorsal setae 27
- Facial ridges with bare. Mid tibia not more than 2 anterodorsal setae.............28
27. Ocellar setae very strong. Gena and pleural region with yellow hairs. Abdomen tawny reddish or reddish yellow........................ *Frontina* Meig.
- Ocellar setae weakk. Gena and pleural region with black hairs. Abdomen tawny black and yellow pollinosit .... undetermined genus and species
28. Four sternopleural setae 29
- Three sternopleural setae...32
29. Facial ridge setose.. *Prosopodopsis* Town.
- Facial ridge bare...30

30. Upper occiput with black setulae behind the postocular row. Palpi normal .... *Bactromyia* B.B.
- Upper occiput without black setulae behind the postocular row. Palpi slender 31
31. Mid tibia with one isolated anterodoral setae. Ocellar setae present. Abdominal t_{1+2} with marginal setae .... *Erycia* Bar.
- Mid tibia with two isolated anterodoral setae. Ocellar setae absent. Abdominal t_{1+2} without marginal setae .... *Alsomyia* Vill.
32. Legs reddish yellow. Hind tibia with preapical posterodorsal setae........ 33
- Legs brownish black. Hind tibia with or without preapical posterodorsal setae 34
33. Facial ridge easily seen in profile. Second costal sector haired ventrally. Basicosta orange yellow .... *Erythrocera* R.D
- Facial ridge not easily seen in profile. Second costal sector bared ventrally. Basicosta brownish black........ *Erycia* Bar.
34. Second costal sector haired ventrally. Upper occiput with black occipital setulae behind the postocular row 35
- Second costal sector bare ventrally. Upper occiput with or without black occipital setulae behind the postocular row 37
35. Hind tibia without preapical posterodorsal setae........ 36
- Hind tibia with strong preapical posterodorsal setae .... *Eurysthaea* R.D.
36. Facial ridge bare. Ocellar setae present .... *Bactromyia* B.B
- Facial ridge setulose. Ocellar setae absent .... *Elodia* R. D
37. Head with two pairs of reclinate orbital setae........ 38
- Head with one pairs of reclinate orbital setae 44
38. Facial ridge setose on at least half its length. Upper occiput without black occipital setulae behind the postocular row 39
- Facial ridge bare or with small hairs. Upper occiput with fine black occipital setulae behind the postocular row 41
39. Hind tibia with preapical anterodorsal setae 40
- Hind tibia without preapical anterodorsal setae .... *Prosopodopsis* Town.
40. Abdominal t_{1+2} with marginal setae. Mid tibia with 3 strong anterodorsal setae........ *Lydellina* Vill.
- Abdominal t_{1+2} without marginal setae. Mid tibia with 1 strong anterodorsal setae........ *Prosopodopsis* Town.

41. Antennal axis far above the level of eye middle.
    Antennae bright orange .... *Bactromyia* B.B
- Antennal axis slightly above the level of eye middle.
    Antennae blackish or dark brownish.. 42

42. Abdominal t_3 and t_4 with median discal setae.
    2 sterno pleural setae. Barette haired on its anterior half .... *Scaphimyia* Mesn.
- Abdominal t_3 and t_4 without median discal setae. 3 sterno
    pleural setae. Barette bare or with small hairs on its anterior half.................... 43

43. Lower ends of facial ridges visible. Gena wider than
    3rd antennal segment. Basicosta yellow .... *Xylotachina* B.B
- Lower ends of facial ridges invisible. Gena slightly narrower
    than 3 rd antennal segment. Basicosta dark brown or blackish .... *Erycia* Bar.

44. Profrons extremely prominent. Abdominal t_3 and t_4 with
    pair of median discal setae .... *Metoposisyrops* Town.
- Profrons less prominent. Abdominal t_3 and t_4 without
    pair of median discal setae .... *Cassidophaga* Bar.

Genus Aneogmena Brauer and Bergenstamm (1891)

The genus *Aneogmena* is raised by Brauer and Bergenstamm (1891). Four species have been reported from world but no species have been reported from India. The genus shows following characters:

1. Abdomen without discal setae on 3rd and 4th tergite.
2. Facial ridges bare except few hairs on setulae above vibrissae.
3. Basal node of r_{4+5} with a row of 3 or more setulae.
4. Eyes bare.

***Anaeogmena striatalis* sp.n.**

(Plate 17, Figures 136 to 144)

Male (Figures 136-144)

8.9mm in length, 0.37mm in width; head 1.76 mm in length, 0.88mm in width; thorax 3.69mm in length, 2.55mm in width; abdomen 5.21mm in length, 1.97mm in width; wing 7.66mmin length, 2.93mm width; halter 1.48mm in length, 0.41mm in width.

Head (Figures 138 and 139)

1.76 mm in length, 0.88mm in width; vertex narrow; inner vertical bristles with single reclinate pair; outer vertical bristle single pair, 2 proclinate orbital setae weak; ocelli present; ocellar bristles short, proclinate; frontal vita dark brown; frontal bristles single row medioclinate; face concave silvery white; lower facial margin protruded forward; vibrissa small weak; facial ridge slightly concave; parafacial silvery white; gena with small hairs; genal dilation well developed; back of head convex with few hairs. **Eyes** 0.93mm in length, 1.21mm in width; bare. **Antenna** (Figure 139) silvery-light brown with scape short, 0. 07mm in length, 0.11mm in width; pedicel short, 0.09mm, 0.13mm in width with few hairs, first flagellomere 0.73mm in length, 0.13mm in width, antennal third segment more than thrice as long as second segment. **Arista** pubescent. **Mouth** proboscis well developed; prementum short with labella; palpus filiform parallel sided, brownish.

Antennal Formula

S L/W=0.636, P L/W=0.693, F L/W=5.615, A=2.314.

Thorax (Figures 140 and 141)

3.69mm in length, 2.55mm in width; humeral callus 3 setae dorsally projected; proepisternum bare; proepimeron pair of setae present. **Scutum** 1.89 in length, 2.01mm in width; acrostichal bristles strong, reclinate (presutural 3+ postsutural 3); dorsal central bristles strong; reclinate (presutural 3+ postsutural 3); notopleuron with 2 setae; postalar callus with 2 setae. **Scutellum** 1.8mm in length, 1.34mm in width; basal scutellar bristles divergent; subapical scutellar bristles very long,

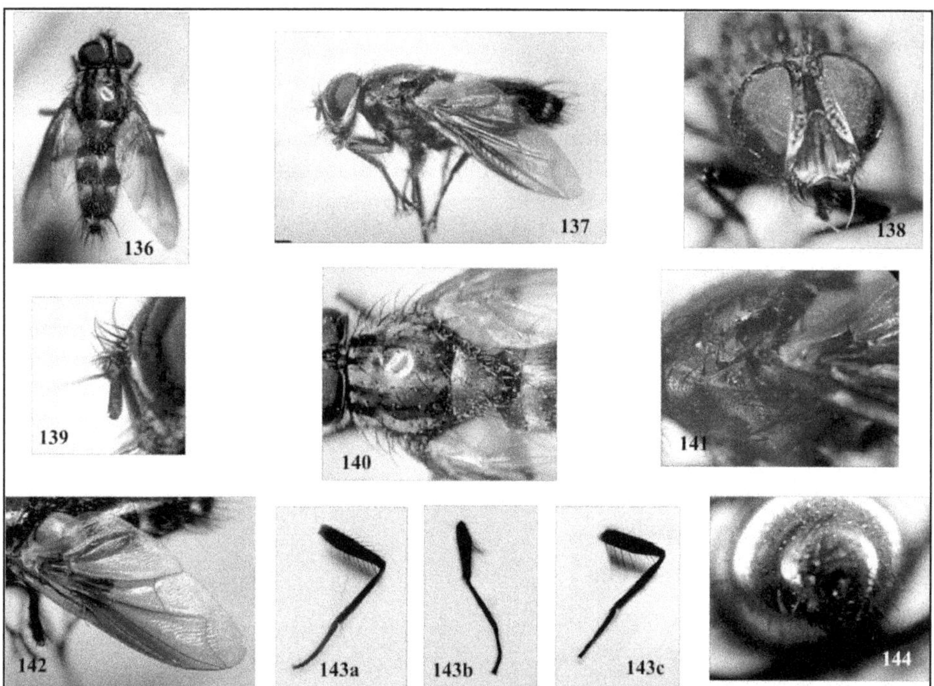

Plate 17. Figure 136: *Anaeogmena striatalis* sp.n. (Male) Dorsal view; Figure 137: Lateral View; Figure 138: Frontal View of Head; Figure 139: Antenna; Figure 140: Dorsal View of Thorax; Figure 141: Lateral View of Thorax; Figure 142: Fore Wing; Figure 143a: Fore Leg; Figure 143b: Mid Leg; Figure 143c: Hind Leg; Figure 144: Genitalia.

divergent. **Subscutellum** convex; katepisternum with 3 strong setae; anepimeron with hairs; katepimeron bare; anterior thoracic spiracle narrow, with closed fringes of hairs.

Wing (Figure 142)

7.66mmin length, 2.93mm width, transparent with very faint brownish shade at proximal end; lower calypter small slightly separated from scutellum; second costal section bare; base of costa with long costal bristle; costigial setae long, strong; fourth costal section longer than 6th costal section; r_1 bare; cu a_1 bare; bend of m distinct and shows small projection of m_2; wing cell r_{4+5} open; cross vein r-m present; cross vein d m-cu present; anal vein not reaching towards hind margin of wing. **Halter** reddish yellow, 1.48mm in length, 0.41mm in width.

Leg

Fore and mid femora with row of spines ventrally. **Forelegs (**Figure 143a) femora 3.53mm; tibia 3.34mm; tarsus 3.62mm; with 1 preapical anterodorsal seta, 1 preapical anteroventral seta. **Mid legs** (Figure 143b) femora 3.56mm; tibia 3.30mm; tarsus 3.56mm; mid tibia 1 preapical anteriodorsal setae present, 1 preapical anteroventral seta, 1 anterodorsal seta. **Hind legs** (Figure 143c) femora 3.60mm; tibia 3.23mm; tarsus 3.61mm; hind tibia,1 preapical posteroventral seta present, 2 anteriodorsal bristles present, 1 preapical anteroventral seta; hind coxa with 1 short seta.

Abdomen (Figures 136 and 137)

5.21mm in length, 1.97mm, elongated, black with dusting in each tergites; abdominal tergite 3rd, 4th, 5th with one pair of erect marginal setae. abdominal tergite 4th slightly longer than tergite 5th; mid dorsal depression on syntergite 1+2 reaching of that segment; tergite 5th tapered at end.

Genitalia (Figure 144)

Terminalia retracted within abdominal tergite 5; tergite 6 reduced joining, segment 7, abdominal tergite 7th and 8th joined to form syntergite 7+8. cercus and surstylus present below syntergite 7+8.

Colour

Black: Head, thorax, abdomen, fore tibia, mid tibia, hind tibia.

Brown: Antenna, palpus.

Dark brown: Frontal vita.

Silvery white: Face, parafacial.

Reddish brown: Eyes.

Host- Unknown.

Host Plant- *Solidago canadensis, Rorippa indica.*

Holotype- Male, India, Maharashtra, Coll. 8-VIII-2013, Shahuwadi, A.S. Desai; legs on slides, labeled as above.

Paratype- 7 Males, Sex ratio (M: F) 7:0, July 2011- February 2015; legs on slides, rest part pinned, preserved in insect box.

Etymology- The species name *striatalis* is described as per the colour 'striations' which is typical characteristic found.

Distributional Record

India, Maharashtra, ♂ 2 Panhala 13-VIII- 2013; ♂3 Shahuwadi 08-XI-2013; ♂ 1 Jath 10-XI- 2013; ♂ 1 Kudal 16- XI-2013.

Remarks

According to key of Crosskey (1976). No species under this genus have been described from India this is first record of this genus. It is having following characters,

1. Wing shows typical shade in different striations.
2. Palpi black.
3. Dusting is present on anterior half of each tergite.
4. Two sternopleural setae.
5. Antennal formula: S L/W=0.636, P L/W=0.693, F L/W=5.615, A=2.314.

Key to Tribes of the Sub-family–Dexiinae

The sub-family Dexiinae was erected by Sabrosky and Arnaud (1965) and Crosskey (1973, 1973) first time reported it as proseninae but Dexiinae is the correct name under the rules of nomenclature. According to Crosskey (1976), 3 tribes and near about 12 genera have been reported from oriental region under this genus. Key to tribes of sub-family tachininae is as follows:

1. Postalar callus with 4 or more strong setae. Scutellum with at least four pairs of marginal setae. Propleuron haired. Various form with metallic color.......... Rutiliini

– Postalar callus with two strong setae. Scutellum with two to three pairs of marginal setae. Propleuron haired or bare. Various form without metallic color.......... 2

2. Head profile subtriangular, profrons prominent. Abdominal t_{4+5} excavate at only its base. Scutellum with two pairs of strong marginal setae. Facial carina absent.......... Doleschallini

– Head profile conspicuously subtriangular, profrons slightly weak. Abdominal t_{4+5} excavate at its hind margin. Scutellum with three pairs of strong marginal setae. Facial carina absent or present.......... **Dexiini**

Tribe Dexiini

The Tribe Dexiini was studied by van Meigen (1826) from oriental region. According to Crosskey (1976), near about 10 genera have been reported from oriental region.

Key to Genera of Tribe Dexiini

1. Proboscis short. Pteropleural seta differentiated. Palpi well developed, longer than third antennal segment 2
– Proboscis very, long and slender. Pteropleural seta absent. Palpi reduced, shorter than third antennal segment .... ***Prosena*** Pel. and Sev.
2. Abdomen with t $_{1+2}$ excavate to its hind margin. Palpi normal 3
– Abdomen with t $_{1+2}$ not excavate to its hind margin. Palpi very strong clubbed .... *Dexiotrix* Vill.
3. Wing cell r_5 open or closed. Pleural regions of thorax with black or brownish black hair. Bend of vein m nearer to m-cu than to the wing margin 4
– Wing cell r_5 closed. Pleural regions of thorax with white or yellowish hair. Bend of vein m nearer to m-cu than to the wing margin .... *Dolichodexia* B.B
4. Second costal sector of wing haired ventrally 5
– Second costal sector of wing bare ventrally 6
5. Propleuron bare. Head with facial carina. Palpi not flattened .... ***Dexia*** Meig.
– Propleuron hairy. Head without facial carina. Palpi flattened .... (Undetermined genus)
6. Two presutural setae. Pre alar absent. Very small forms .... *Tylodexia* Town.
– Three or more presutural setae. Pre alar present. Larger forms 7
7. Propleuron bare. Head with facial carina .... *Myostoma* R.D
– Propleuron haired. Head with or without facial carina 8
8. Three postdorsocentral setae. Abdomen with t $_{1+2}$ with or without pair of strong marginal setae 9
– Four postdorsocentral setae. Abdomen with t $_{1+2}$ without pair of strong marginal setae .... *Billea* R.D
9. Abdominal t_3 and t_4 without discal setae. Profrontal region of head bare 10
– Abdominal t_3 and t_4 with discal setae. Profrontal region of head with black hairs .... *Dinera* R. D
10. Abdominal tergite 5th of male produced to sharp posterior point with spiniform setae .... *Urodexiomima* Town.
– Abdominal tergite 5th of male not produced to sharp posterior point with spiniform setae .... *Philippodexia* Town.

Genus – *Prosena* Le peletier and Serville (1828)

Genus *Prosena* is raised by Le peletier and Serville (1828). According to key of Rayner and Raper (2001) only one species is reported from the world. The genus shows following characters:

1. Proboscis very long and slender than the height of head.
2. Pteropleural seta absent.
3. Palpi much reduced than third antennal segment.

Key to Species of the Genus *Prosena* Le peletier and Serville (1828)

1. Tergite 2nd hollowed dorsally to the posterior edge.
   Only one intra alar present...***siberita*** F.

Prosena siberita (Fabricus, 1775)

(Plate 18, Figures 145 to 153)

Male (Figures 145-153)

Body 9.92 mm in length, 3.55 mm in width; wing 8.18mm in length, 2.73mm in width; antenna 0.96 mm long; halter 1.21mm in length, 0.39mm in width.

Head (Figures 147 and 148)

1.36 mm in length, 3.11mm in width; vertex narrow; inner vertical bristles one pair, reclinate; outer vertical bristles absent; ocelli present; ocellar bristles small proclinate; frontal vitta well developed, dark orange-red; frontal bristles reach down to the base of 1st antennal segment; fronto orbital plate silver white; reclinate absent; proclinate orbital bristles two pairs; face less concave; lower facial margin protruding forward clearly visible in lateral view; vibrissa strong; facial ridge concave; parafacial bare; gena narrow; genal dilation well developed; back of head concave, with dense hairs. **Eyes** 0.74mm in length, 2.54mm in width, bare, red. **Antenna** (Figure 148) scape 0.16mm in length, 0.11mm in width; pedicel 0.24mm in length, 0.16mm in width; first flagellomere 0.56mm in length, 0.13mm in width, concave. **Arista** plumose; proboscis very long and slender, its length conspicuously greater than height of head. **Mouth** prementum elongated with short labella; palpus reduced.

Antennal Formula

S L/W=1.454, P L/W=1.5, F L/W=4.307, A= 2.420.

Thorax (Figures 149 and 150)

4.20mm in length, 3.23mm in width; hairs on thorax yellowish; humeral callus with reclinate 3 setae; proepisternum bare; proepimeron with 2 strong setae dorsally projected; prosternum with some hairs. **Scutum** 2.90 mm in length, 3.17mm in width; acrostichal bristles (presutural 3+ postsutural 4); dorsal central bristles (presutural 3+ postsutural 4); notopleuron with 2 strong, reclinate setae; one intra alar setae present; postalar callus with 2 strong reclinate setae. **Scutellum** 1.30 mm in length,

1.71mm in width; basal scutellar bristles strong, long; lateral scutellar bristle weak; subapical scutellar bristles long, strong; apical scutellar bristles parallel, strong. **Subscutellum** convex, bare; katepisternum 2 strong setae; anepimeron without pteropleural setae; katepimeron with pale yellow dense hairs; meron with vertical row of small hairs; anatergite with small hairs; anterior thoracic spiracle narrow and closed by fringes of hairs.

Wing (Figure 151)

8.18mm in length, 2.73mm in width, transparent; lower calypter well developed, inner margin more contiguous to lateral margin of scutellum; second costal section bare; base of costa with small hairs; costal bristles absent; fourth costal section longer than 6th costal section; r_1 bare; cu a_1 bare; bend of m obtuse; wing cell r_{4+5} open; cross vein r-m present; cross vein d m-cu present; anal vein not reaches towards hind margin of wing. **Halter** reddish yellow, 1.21mm in length, 0.39mm in width.

Leg

Fore and mid femora with double rows of spines at ventrally. **Fore legs** (Figure 152a) femora 2.78mm; tibia 2.24mm; tarsus 3.24mm; fore tibia with one preapical anteriodorsal setae, fore tarsus laterally compressed. **Mid legs** (Figure 152b) femora 2.75mm; tibia 2.19mm; tarsus 3.21mm; mid tibia with one anteriodorsal setae. **Hind**

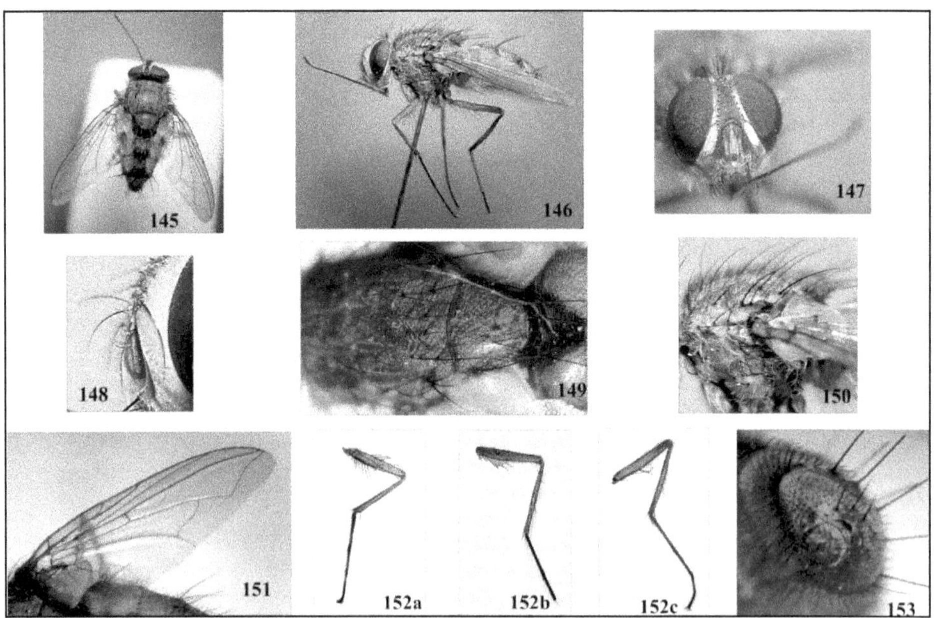

Plate 18. Figure 145: *Prosena siberita* (Fabricus) (Male) Dorsal view; Figure 146: Lateral View; Figure 147: Frontal View of Head; Figure 148: Antenna; Figure 149: Dorsal View of Thorax; Figure 150: Lateral View of Thorax; Figure 151: Fore Wing; Figure 152a: Fore Leg; Figure 152b: Mid Leg; Figure 152c: Hind Leg; Figure 153: Genitalia.

legs (Figure 152c) femora 2.73mm; tibia 2.20mm; tarsus 3.18mm; hind tibia with one preapical posterioventral setae; hind coxa with 1 seta.

Abdomen (Figures 145 and 146)

4.98 mm in length,3.22 mm in width, without discal setae, even first segment normally with strong marginals; abdominal tergite 2th hollowed ventrally; abdominal tergite 5th as long as 4th tergite; mid dorsal depression 1+2 syntergite reaches up to 2 tergite margin; discal bristles present; abdominal sternites overlapped by ventral edges of tergites.

Genitalia (Figure 153)

Male terminalia retracted within abdominal tergite 5; tergite 6th less reduced, joining segment 7th and dorsoventrally slightly flattened.

Colour

Dark brown: Frontal vita.

Yellow: Fore tibia, mid tibia, hind tibia, thorax, abdomen, antenna.

Silvery white: Face, parafacial, head, palpus.

Reddish brown: Eyes.

Host- unknown.

Host Plant- Bark of Mango tree, Eucalyptus, *Helianthus elastic.*

Paratype- 10 Males, Sex ratio (M: F) 10:0, July 2011- February 2015; legs on slides, rest part pinned, preserved in insect box.

Distributional Record

India, Maharashtra, ♂3 Panhala 16- VII-2013; ♂2 Amboli 21-XI-2013; ♂1 Kudal 22-XI- 2013; ♂1 Vita 08-XI- 2013; ♂ 1 Wai 11-X- 2014; ♂ 2 Karveer 03-VII- 2015.

Remarks

According to key of Crosskey (1976) and Rayner and Raper (2001) this species is *Prosena siberita* (Fabricus, 1775). Following characters are similar according to above key:

1. Proboscis very long and slender than the height of head.
2. Pteropleural seta absent.
3. Palpi much reduced than third antennal segment.

Additional characters are as follows:

1. Lateral region of thorax shows Pale brownish hairs present.
2. Frons 2.5x as wide as one eye.
3. **Antennal formula:** S L/W=1.454, P L/W=1.5, F L/W=4.307, A= 2.420.

Genus – *Dexia* Meigen (1826)

Genus *Dexia* was visualized by Meigen (1826). No species from India have been reported under this genus. According to key of Rayner and Raper (2001) two species are reported from world. The genus shows following characters,

1. Propleuron bare.
2. Head with facial carina.
3. Palpi present but not flattened.
4. Humeral callus 2 to 3 bristles.
5. Cell r_5 open.

Key to Species of the Genus *Dexia* Meigen (1826)

1. Three to four sternopleural setae. Abdomen dusted to the end in males with yellow basic color with black central stripe, in female's dark brown .... *rustica* Fall.
– Two sternopleural setae 2
2. Abdomenal tergites in females with more or less black color .... ***vacua*** Fall.
– Abdomenal tergites in females completely black color .... ***indica*** sp. n.

Dexia vacua (Fallen, 1816)

(Plate 19, Figures 154 to 162)

Male (Figures 154-162)

Body 10.46 mm long, 3.85 mm broad; 9.06mm in length, 3.22mm in width; antenna 0.77 mm long; halter 1.16mm long, 0.42mm wide.

Head (Figures 156 and 157)

1.18mm in length, 2.40mm in width; vertex narrow; inner vertical bristles single pair, strong, reclinate; outer vertical bristle single pair, reclinate; ocelli present; ocellar bristles single pair, proclinate; frontal vitta well developed; frontal bristles medioclinate; fronto orbital plate with one row medioclinate setae; reclinate upper orbital bristles absent; proclinate orbital bristles absent; face less concave, dark brown; lower facial margin seen in lateral view; vibrissa strong with small setulae at base; facial ridge convex; parafacial bare; gena narrow; genal dilation with well developed three strong setae; back of head slightly convex with small hairs. **Eyes** 0.86mm in length, 1.63mm in width, bare, reddish brown. **Antenna** (Figure 157) scape 0.11mm in length, 0.10mm in width, very short, pedicel short, 0.18mm in length, 0.15mm in width; first flagellomere 0.48mm in length, 0.14mm in width, yellow, short, straight, pointed at tip. **Arista** plumose type, arising from base of first flagellomere; **Mouth** proboscis well developed; prementum slightly elongated with short labella; palpus well developed slightly swollen at apex, not flattened.

Antennal Formula

S L/W=1.1, P L/W=1.2, F L/W=3.428, A=1.909.

Thorax (Figures 158 and 159)

3.46mm in length, 2.54mm in width, humeral callus with 2 setae; propleuron bare; proepimeron with 1 propleural setae, 1 prostigmatic seta; prosternum bare. **Scutum** 2.76mm in length, 2.51mm in width, 4 dark longitudinal strips; acrostichal bristles very small (presutural 1 + postsutural 1); dorsal central bristles small (presutural 3+ postsutural 3); notopleuron 2 strong setae; postalar 2 strong setae. **Scutellum** 0.7mm in length, 1.48mm in width; basal scutellar bristles strong, lateral scutellar bristles absent; subapical scutellar bristles more stronger than other bristles; apical scutellum bristles crossed, strong. **Subscutellum** convex; katepisternum (sternopleuron) 2 strong sternopleural setae; anepimeron with bristles at the base of wing; katepimeron bare; meron with vertical row of bristles; postmetacoxal bare; anatergite bare; anterior thoracic spiracle narrow and closed by hairs.

Wing (Figure 160)

9.06mm in length, 3.22mm in width, transparent; lower calypter well developed with hairs at outer edge; second costal section with fine hairs at margin; base of costa without strong bristles; costal bristles present; fourth coastal section longer than 6th costal section; r_1 bare; cu a_1 bare; bend of m distinct; wing cell r_{4+5} open at wing

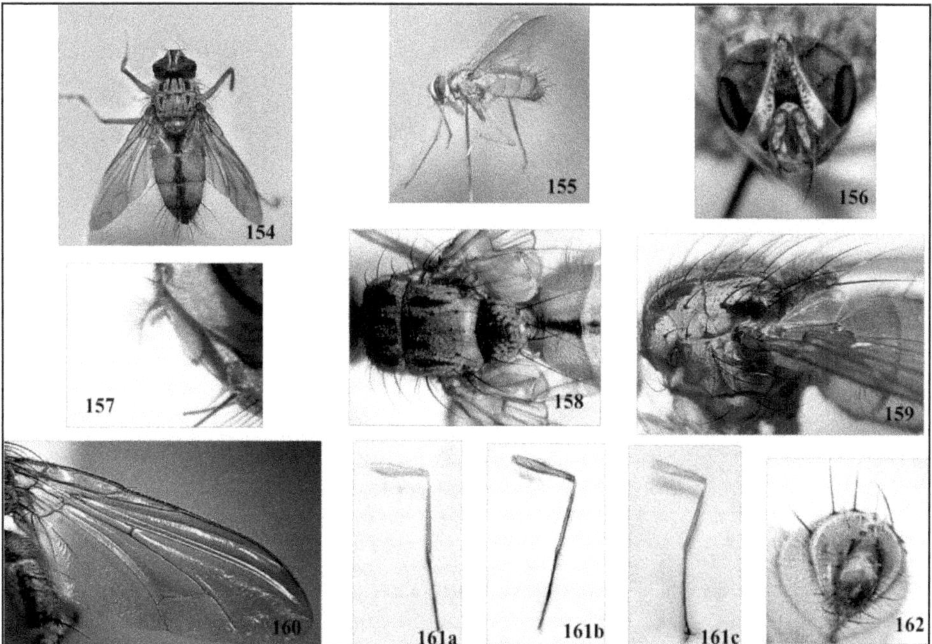

Plate 19. Figure 154: *Dexia vacua* Fallen (Male) Dorsal view; Figure 155: Lateral View; Figure 156: Frontal View of Head; Figure 157: Antenna; Figure 158: Dorsal View of Thorax; Figure 159: Lateral View of Thorax; Figure 160: Fore Wing; Figure 161a: Fore Leg; Figure 161b: Mid Leg; Figure 161c: Hind Leg; Figure 162: Genitalia.

margin; cross vein r-m present; cross vein d m-cu distinctly oblique; anal vein not reaching hind margin of wing; **Halter** dark yellow, 1.16mm length, 0.42mm width.

Leg

Fore and mid with row spine; **Forelegs** (Figure 161a) femora 4.74mm; tibia 3.74mm; tarsus 4.60mm; fore tibia yellow, one long preapical anteriodorsal setae; fore tarsus laterally compressed, hard; fore claws present; **Mid legs** (Figure 161b) femora 4.69mm; tibia 3.55mm; tarsus 4.58mm; mid tibia yellow, with 2 anterio dorsal setae; **Hind legs** (Figure 161c) femora 4.72mm; tibia 3.64mm; tarsus 4.59mm; mid tibia yellow, with 1 anterodorsal setae; hind coxa with 2 setae.

Abdomen (Figures 154 and 155)

5.36mm in length, 3.31mm in width, oval shaped, yellow with less black posterior edge of 4th and 5th tergites; tergite 2 dorsally hollowed to posterior edge; abdominal tergite 5th as long as 4th tergite; mid dorsal depression reaches back to hind margin of that segment; discal bristles present 3 but absent on syntergite 1+2, abdominal sternite overlapped by the ventral edges of tergite.

Genitalia (Figure 162)

Terminalia retracted within abdominal tergite 5; tergite 6 reduced; surstylus, cercus hidden inside epandrium.

Colour

Yellow: Fore tibia, mid tibia, hind tibia, thorax, abdomen, palpus.

Silvery white: Face, parafacial, head, antenna.

Dark brown: Frontal vita.

Reddish brown: Eyes.

Host- *Dacus carota, Tecoma castanifolia, Helianthus elastic, Aster amellus.*

Host Plant- Unknown.

Paratype- 10 Males, Sex ratio (M: F) 10:0, July 2011- February 2015; legs on slides, rest part pinned, preserved in insect box.

Distributional Record

India, Maharashtra, ♂3 Panhala 13-VIII- 2013; ♂2 Hatakanangale 28- XI-2013; ♂1 Shahuwadi 08-XI-2013; ♂1 Gargoti16-VII-2013; ♂ 1 Phaltan 02-XI- 2014; ♂ 2 Mahabaleshwar 09-XI- 2014.

Remarks

According to key of Crosskey (1976) and Rayner and Raper (2001) this species is *Dexia vacua* (Fallen, 1816), Following characters are similar according to above key:

1. 2 sternopleural setae.
2. Abdomen tergite 4th and 5th in males more or less black at posterior edges.
3. Dusting of tergite covers 1/2- 5/6 segment length.

Additional characters are as follows:

1. Frons 2.55x as wide as one eye.
2. Bend of m shows slight presence of m_2.
3. Third antennal segment is 2.5 -3x as wide as Second antennal segment.
4. Last medioclinate frontal setae near to base of 1st antennal segment.
5. **Antennal formula:** S L/W=1.1, P L/W=1.2, F L/W=3.428, A=1.909.

Dexia indica sp.n.

(Plate 20, Figures 163 to 171)

Female (Figures 163-171)

Body 12.48 mm in length, 4.15 mm in width; wing 10.96mm in length,3.56mm in width; antenna 0.96 mm long; halter 1.36mm in length, 0.49mm in width.

Head (Figures 165 and 166)

1.7mm in length, 3.01mm in width; yellowish brown with strong facial carina. vertex wide; inner vertical bristles reclinate; outer vertical bristle absent; ocelli present; ocellar bristles short, weak, proclinate; frontal vitta well developed; frontal bristles medioclinate; fronto orbital plate with one additional row medioclinate setae; reclinate upper orbital bristles absent; proclinate orbital bristles with two setae; face concave, dark brown; lower facial margin less visible in lateral view; facial carina present; vibrissa strong, crossed; facial ridge convex; parafacial bare; gena narrow; genal dilation with well developed; back of head convex, less hairs pale. **Eyes** 0.74mm in length, 2.41mm in width, bare, reddish brown; **Antenna** (Figure 166) scape 0.13mm in length, 0.09 in width, very short, silvery yellow, pedicel 0.21mm in length, 0.11mm in width; first flagellomere 0.68mm in length, 0.12mm in width. **Arista** plumose type, initiated from base of first flagellomere; **Mouth** proboscis well developed; prementum with short labella; palpus well developed slightly swollen at apex.

Antennal Formula

S L/W=1.45, P L/W=1.9, F L/W=6.18, A=3.17.

Thorax (Figures 167 and 168)

4.46mm in length, 3.52mm in width; humeral callus 2 setae, reclinate; propleuron bare; proepimeron with 1 propleural setae and 1 prostigmatic seta; prosternum bare. **Scutum** 3.18mm in length, 3.49mm in width, 4 dark longitudinal incomplete strips; acrostichal bristles reclinate (presutural 3 + postsutural 3); dorsal central bristles reclinate (presutural 3+ postsutural 3); notopleuron with 2 reclinate setae; postalar callus with 2 reclinate setae. **Scutellum** 1.28mm in length, 1.98mm in width; basal scutellar bristles single pair; lateral scutellar bristles absent; subapical scutellar bristles one pair, longer and stronger than other bristles; apical scutellum bristles crossed, short, strong. **Subscutellum** convex; katepisternum 2 strong setae (1 reclinate + 1 proclinate); anepimeron with one bristles at the base of wing; katepimeron bare; meron with vertical row of bristles; postmetacoxal bare; anatergite bare; anterior thoracic spiracle narrow and closed by fringes of hairs.

Wing (Figure 169)

10.96mm in length,3.56mm in width, transparent; lower calypter well developed inner margin less contiguous to lateral margin of scutellum; second costal section with fine hairs at ventrally; base of costa without strong bristles; costal bristles absent; fourth coastal section longer than 6th costal section; r_1 bare; cu a_1 bare; bend of m abrupt; wing cell r_{4+5} open at wing margin, hairs at base are absent; cross vein r-m present; cross vein d m-cu distinctly present; anal vein not reaching hind margin of wing. **Halter** reddish yellow, 1.36mm length, 0.49mm width.

Leg

Extremely elongated legs, fore and mid femora with rows of spiny bristles. **Fore legs** (Figure 170a) femora 3.4mm; tibia 3.58mm, tarsus 4.64mm; fore tibia with 1 preapical anteriodorsal setae; fore tarsus laterally compressed, hard. **Mid legs** (Figure 170b) femora 3.38mm; tibia 3.52mm, tarsus 4.60mm; mid tibia with 2

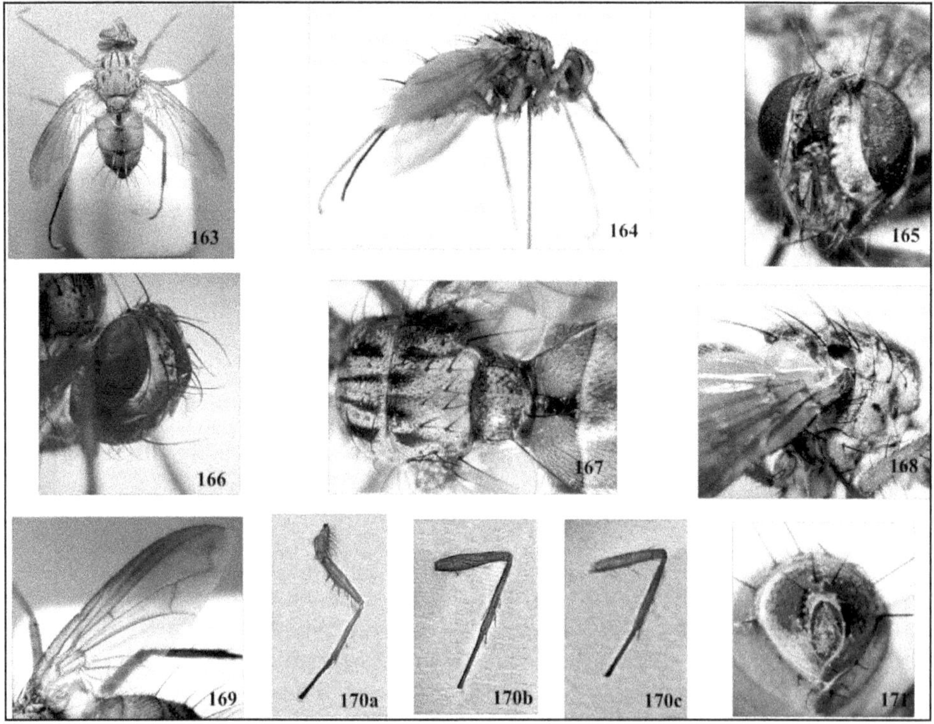

Plate 20. Figure 163: *Dexia indica* sp.n. (Female) Dorsal view; Figure 164: Lateral View; Figure 165: Frontal View of Head; Figure 166: Antennae; Figure 167: Dorsal View of Thorax; Figure 168: Lateral View of Thorax; Figure 169: Fore Wing; Figure 170a: Fore Leg; Figure 170b: Mid Leg; Figure 170c: Hind Leg; Figure 171: Genitalia.

anteriodorsal setae. **Hind legs** (Figure 170c) femora 3.42mm; tibia 3.60mm, tarsus 4.61mm;hind tibia with 1 anterodorsal setae; hind coxa with 2 small setae.

Abdomen (Figures 163 and 164)

6.32mm in length, 3.09mm in width, oval shaped; abdominal tergite 5th as long as 4th tergite; mid dorsal depression present reaches back to hind margin of that segment; median discal bristles present, abdominal sternite overlapped by the ventral edges of tergite.

Genitalia (Figure 171)

Terminalia retracted within abdominal 5th tergite; tergite 6th less reduced, joining segment 7th; prominent parts of genitalia are piercer, cercus: both parts turn inward and hidden.

Colour

Dark brown: Frontal vita.

Yellow: Fore tibia, mid tibia, hind tibia, abdomen.

Golden yellow: Face, parafacial, antenna.

Reddish brown: Eyes, head, thorax.

Host- Unknown.

Host plant- *Chrysanthemum* sp.

Holotype- Female, India, Maharashtra, Coll. 23- IX-2012, Barki, A.S. Desai; antenna, wings, legs, tergites on slides, labeled as above.

Paratype- 5 Females, Sex ratio (M: F) 0:5, July 2011- February 2015; legs on slides, rest part pinned, preserved in insect box.

Etymology- The species has been reported first time from India hence the species name *indica.*

Distributional Record

India, Maharashtra, ♀ 2 Kagal 08- VII-2013; ♀ 1 Panhala 15-VII- 2013; ♀ 1 Radhanagari 16- VII-2013; ♀1 Shahuwadi 13-X-2013.

Remarks

According to key of Rayner and Raper (2001) this species runs close to *Dexia vacua* by having following characters:

1. Sternopleuron with 2 sternopleural setae.
2. Calyptrae at outer edge with hairs.
3. Tergites dusted.

However, it differs from the above species by following distinct characters:

1. 4th and 5th abdominal tergites almost completely black.
2. Apical bristles parallel but slightly crossed at tip.

3. Costal spine reduced.
4. Bend M in wings without M_2 appendix.
5. Antennal formula: S L/W=1.45, P L/W=1.9, F L/W=6.18, A=3.17.

Sub-family–Phasiinae

The sub-family Phasiinae was studied by Crosskey (1976). According to Crosskey (1973), 4 tribes and 17 genera have been reported from oriental region.

Key to Tribes of Sub-family Phasiinae

1. Face formed into a swollen carina that extends from the antennae to the epistome and has a sharp median edge. Three sterno pleural setae. Four or five post dorsocentral setae .... Eutherini
- Face not carinate without sharp median edge. One to three post dorsocentral setae 2
2. Abdominal vestiture clearly with setae. One or two post dorsocentral. 1 post intra alar seta or none. Head without proclinate orbital setae .... Phasiini
- Abdominal vestiture hair like. Three post dorsocentral setae. 1 or 2 post intra alar seta or none. Head with proclinate orbital setae 3
3. Abdominal base and hind coxa widely separated. Abdomen very elongate and subcylindrical. Lower calypter rounded on its hind margin. One or two post intra setae. Scutellum with 2 or 3 pairs of marginal setae .... **Cylindromyiini**
- Abdominal base and hind coxa normal. Abdomen subovate. Lower calypter very large. Two strong post intra setae. Scutellum with 3 or 4 pairs of marginal setae .... Leucostomatini

Tribe Cylindromyiini

The tribe Cylindromyiini was studied by Towensend (1936; 1938) from oriental region. According to Crosskey (1976), 7 genera have been reported from oriental region.

Key to Genera of Tribe Cylindromyiini

1. Palpi absent. Cell r_5 with developed petiole. Propleuron bare or haired. Some abdominal terigtes without discal setae 2
- Palpi present. Cell r_5 with or without developed petiole. Propleuron bare. Some abdominal terigtes with discal setae 3
2. Abdomen almost strongly clavate. Basal of r_{4+5} with long setae.. *Gerocyptera* Town.
- Abdomen subcylindrical or subfusiform. Basal of r_{4+5} with tiny setae .... ***Cylindromyia*** Meig.

3. Arista thickened on more than half of its length. Abdomen appearing trisegmented when seen from above .... *Catapariprosopa* Town.

– Arista very long and slender. Abdomen appearing quadri-segmented when seen from above 4

4. Abdominal tergite without median discal setae. Hind tibia with preapical posteroventral seta.Vibrissae very strong and crossed .... *Lophosia* Meig.

– Abdominal tergite 3rd and 4th with median discal setae. Hind tibia without preapical posteroventral seta. Vibrissae very weak and hair like 5

5. Vibrissae strong and almost crossed. Hind tibia 3-4 anterodorsal setae and 3-4 posterodorsal setae.Lower calypter small, flat and subcircular. Wing smoky in color at some regions .... *Penthosiosoma* Town.

– Vibrissae hair like and not crossed. Hind tibia with not more than 2 anterodorsal setae and 2 posterodorsal setae. Lower calypter slightly elongated. Wing smoky in color at some regions 6

6. Abdomen clavate. Vespid mimic .... *Formicophania* Town.

– Abdomen subfusiform. Without vespid mimic .... *Hermya* R.D

Genus–*Cylindromyia* Meigen (1803)

The genus *Cylindromyia* is raised by Meigen (1803). According Gilasian et. al, (2013) from various parts of world more than 18 species of this genus have been reported. The genus shows following characters:

1. Abdomen subcylindrical or subfusiform.
2. Wing cell R_{4+5} closed and also with tiny hairs at its basal node.
3. Terigte and sternite 7th in females as hook like complex.

Key to Species of the Genus *Cylindromyia* Meigen (1803)

1. Hind tibia without posteroventral seta 2

– Hind tibia with 1–2 posteroventral setae 14

2. Vibrissa weak, at most half times as long as face 3

– Vibrissa stronger, more than half times as long as face 9

3. Proepisternum with setulae. Outer marginal setulae of lower calypter long, 2–3 times as long as width of its thickened margin; piercer absent in female 4

– Proepisternum bare; outer marginal setulae of lower calypter at most as long as width of its thickened margin; piercer present in female 5

4. Lateral (outer) vertical setae present. Proepisternum with black setulae or bare. Tegula yellowish-brown. Legs mostly yellowish brown, tarsi black. Female abdominal tergite 3 with ventral spinulae on each side, tergite 4 without posteroventral spinulae; female tergite 5 with median discal setae; male postabdomen .... *persica* Tschorsnig

– Lateral vertical setae absent; proepisternum with pale setulae; tegula black; legs black; female abdominal tergite 3 without ventral spinulae, tergite 4 with posteroventral spinulae; female tergite 5 without discal setae; male sternite 5. .... *rufifrons* Loew.

5. Scutellum brown .... ***scutellata*** sp.n.

– Head, mesopleuron, scutellum and legs yellowish-orange 6

6. Apical scutellar setae very short, at most 0.25 times as long as scutellar subapical setae; posterior supra-alar seta absent .... *rubida* Loew

– Head, mesopleuron, scutellum and legs brownish-black; apical scutellar setae 0.4–0.5 times as long as scutellar subapical setae; posterior supra-alar seta present 7

7. Antenna shorter than face, postpedicel 3.3–4.3 times as long as wide. Back of head with only pale setulae; tergite 3 entirely yellowish-orange, without a dorsal median longitudinal black stripe; male postabdomen. *vallicola* Ziegler and .... gilasian.

– Antenna as long as face, postpedicel 5.0–6.5 times as long as wide back of head with a few black setulae dorsally; tergite 3 with a dorsal median longitudinal black stripe; male postabdomen different 8

8. Wing infuscated in anterior narrow portion (between anterior margin and r 2+3). Palpus absent or hardly visible; mid tibia with 1 anterodorsal seta; male sternite 5 .... *theodori* Kugler

– Wing widely infuscated; palpus distinctly visible, 3–5 times as long as wide; mid tibia with 2 anterodorsal setae 9

9. Syntergite 1+2 with median submarginal setae; abdominal tergites 3+4 in females with lateral marginal setae; male sternite 5 .... *gemma* rich Ter.

– Syntergite 1+2 without median submarginal setae; abdominal tergites 3+4 in females without lateral marginal setae; male sternite 5 .... *robusta* Loew

10. Abdominal tergites 4–5 reddish-orange; tergites 2–4 with a pair of dorsal discal setae (sometimes with 1 seta, rarely absent in female); male sternite 5 .... *bicolor* Olivier

– Abdominal tergites 4–5 brownish-black; tergites 2–4 without dorsal discal setae 11

11. Abdominal syntergite 1+2 with 2–4 long setae posteroventrally; postpedicel 1.6–1.8 times as long as pedicel .... *brevicornis* Loew

– Syntergite 1+2 without long setae posteroventrally; postpedicel more than 1.8 times as long as pedicel 12

12. Lateral (outer) vertical setae absent; frons at its narrowest point in males 0.60–0.72 and in females 0.70–0.75 times as wide as an eye viewed dorsally; back of head with 1–4 black setulae dorsally; basicosta yellowish-brown; hind femur and tibia in male with long posterior and ventral setulae; male sternite 5............................ *pilipes* Loew

– Lateral (outer) vertical setae present; frons at its narrowest point in males 0.70–0.82 and in females 0.80–0.92 times as wide as an eye viewed dorsally; back of head with only pale setulae; basicosta brownish-black; hind tibia and femur in males with normal setulae 13

13. Lateral (outer) vertical setae slightly longer than postocular setae; scutellum with grey microtrichosity viewed from behind; mid leg tarsomeres 2+3 in male black; male sternite 5 .... *brassicaria* Fabricius

– Lateral (outer) vertical setae distinctly longer than postocular setae; scutellum without microtrichosity.mid leg tarsomeres 2+3 in male yellowish-brown.. 14

14. Abdominal tergite 3 with a black dorsal strip (rarely without); tergite 4 yellowish-orange with a wide black dorsal strip narrowed anteriorly (sometimes extends to posterolateral third margin of tergite); hind tibia in male without (sometimes with 1 or rarely with 2) setae anteriorly (next to anteroventral setae); male postabdomen as in Figs 6–8, surstylus distinctly hook-shaped .... *uncinata* gilasian, Talebi and Ziegler

– Abdominal tergite 3 (mostly) entirely yellowish-orange; tergite 4 black except anterior third; hind tibia in males with 4–6 (rarely 2–3) distinct setae located anteriorly next to anteroventral setae; male sternite 5 underside of surstylus nearly straight, not hook-shaped... *montana* Kugler

15. Apical scutellar setae absent; 1 katepisternal seta; inner margin of lower calypter black; apical third of arista in male flattened and broadened; male sternite 5.. *pusilla* Meigen

– Apical scutellar setae present; 2 katepisternal setae; inner margin of lower calypter yellowish-brown, arista in male not broadened 16

16. Antenna as long as face, postpedicel 3–5 times as long as wide; vibrissa short and about 0.4 times as long as face; lower facial margin slightly protruded; all femora and tibiae yellowish-orange; abdomen brownish-black; male sternite .... *rufipes* Meigen

- Antenna shorter than face, postpedicel 2.0–3.6 times as long as wide; vibrissa long, more than 0.5 times as long as face; lower facial margin distinctly protruded; all femora and tibiae black; abdominal tergites 2–3 at least laterally and ventrally yellowish-orange........17

17. Proepisternum with black setulae; lateral (outer) vertical setae present; basal scutellar setae distinct, almost as long as subapical scutelar setae; male sternite 5..*crassa* Loew

- Proepisternum bare; lateral (outer) vertical setae absent; basal scutellar setae very short or absent..18

18. Posterior supra-alar seta present; mesonotum bare between dorsocentral and supra-alar setae; mid tibia with a short posterodorsal seta; postpedicel 3.0–3.5 times as long as wide; female abdominal syntergite $_{1+2}$ without ventral spines; male sternite 5 ....*intermedia* Meigen

- Posterior supra-alar seta absent; mesonotum between dorsocentral and supra-alar setae with setulae; mid tibia without posterodorsal seta; postpedicel 1.8–2.5 times as long as wide; female abdominal syntergite $_{1+2}$ with posteroventral spines; male sternite 5...........*auriceps* Meigen

***Cylindromyia scutellata* sp.n.**

(Plate 21, Figures 172 to 180)

Female (Figures 172-180)

Body 7.95 mm long, 2.29 mm broad; wing 6.16mm in length, 2.21mm in width; antenna 2.12 mm long; halter 1.28mm in length, 0.47mm in width.

Head (Figure 174,175) - 1.51 mm in length,1.81mm in width; silvery yellow; vertex wide; inner vertical bristles strong, long reclinate; ocelli present; ocellar bristles weak, proclinate; frontal vitta, present very well developed, with short hairs; frontal bristles medioclinate but short; fronto orbital plate with row of setae with scattered hairs; lateroclinate 3 setae; face concave, without facial carina; lower facial margin protruding; vibrissa not differentiated from hairs on vibrissal angle; facial ridge straight; parafacial bare; gena narrow; genal dilation undeveloped; back of head convex, with hairs. **Eyes** 0.69mm, in length, 1.41mm in width bare, facets not enlarged, wide apart, brownish. **Antenna**(Figure 175) Antenna as long as face;scape very short, 0.09mm in length, 0.11mm in width; pedicel 0.41mm in length, 0.13mm in width; first flagellomere 1.62mm in length,0.15mm in width with small hairs, convex. **Arista** thickened at base. **Mouth** proboscis seldom; prementum short with small labella; palpus absent.

Antennal Formula

S L/W=0.818, P L/W=3.153, F L/W=10.8, A=4.923.

Thorax (Figures 176 and 177)

1.78mm in length, 1.73mm in width orange-red; humeral callus with reclinate 2 setae; proepisternum with very small hairs; proepimeron with 2 setae; prosternum bare. **Scutum** 0.93 mm in length,1.71mm in width with 2 black strips; acrostichal bristles with many setulae; dorsocentral bristles with many setulae; propleuron bare; notopleuron with 2 setae strong, reclinate; posterior supralar seta absent.; postalar callus with strong reclinate 1 setae. **Scutellum** 0.85mm in length,1.17mm in width; lateral scutellar bristle short; subapical scutellar bristles very long, strong; apical scutellar bristles very short than subapical scutellar bristles. **Subscutellum** convex, brownish; katepisternum with 2 setae present; anepimeron bare; katepimeron bare; meron bare; anatergite bare; anterior thoracic spiracle narrow and closed by fringes of hairs.

Wing (Figure 178)

6.16mm in length, 2.21mm in width, with dark pattern, well developed wing; lower calypter well developed, rounded on its hind margin; second costal section with hairs ventrally; base of costa with small hairs; costal bristles not differentiated than costigeal setae; fourth costal section as long as 6th costal section; r_1 bare; cu a_1 bare; bend of m with short stub and right angled; wing cell r_{4+5} open, in case petiole m joining r_{4+5} at an acute angle; basal node of r_{4+5} tiny inconspicuous hairs ventrally;

Plate 21. Figure 142: *Cylindromyia scutellata* sp.n. (Female) Dorsal view; Figure 173: Lateral View; Figure 174: Frontal View of Head; Figure 175: Antennae; Figure 176: Dorsal View of Thorax; Figure 177: Lateral View of Thorax; Figure 178: Fore Wing; Figure 179a: Fore Leg; Figure 179b: Mid Leg; Figure 179c: Hind Leg; Figure 180: Genitalia.

cross vein r-m present; cross vein d m-cu present; anal vein not reaches towards hind margin of wing. **Halter** reddish brown, 1.28mm in length, 0.47mm in width.

Leg

Fore and mid femora with 2 rows of spines at ventrally; **Fore legs** (Figure 179a) femora 2.48mm; tibia 2.31mm and tarsus 2.32mm; foreleg with small preapical anteriodorsal setae, preapical posterioventral seta small. **Mid legs** (Figure 179b) femora 2.37mm; tibia 2.29mm and tarsus 2.25mm, mid tibia with 1 anterodorsal setae; 1 preapical anteriodorsal, 1 preapical posterial ventral. **Hind leg** (Figure 179c) femora 2.49mm; tibia 2.36mm and tarsus 2.39mm hind tibia with 2 long anteriodorsal setae, long preapical anteriodorsal setae, one submedian preapical posterioventral setae; hind coxa with minute hairs.

Abdomen (Figures 172 and 173)

4.66 mm in length,1.29mm in width, subcylindrical, very elongated; abdominal tergite strongly elongated with dusting on each tergite; mid dorsal depression t_{1+2} syntergite reaches to 2 tergite margin; discal bristles absent; abdominal sternites overlapped by ventral edges of tergites.

Genitalia (Figure 180)

Female terminalia with hook like clasper with curved end.

Colour

Brown: Antenna, palpus, thorax, abdomen.

Yellow: Frontal vita.

Yellow orange: Fore tibia, mid tibia, hind tibia.

Silvery yellow: Head, face, parafacial.

Reddish brown: Eyes.

Host- Unknown.

Host Plant- *Caltropis* sp.

Holotype- Male, India, Maharashtra, Coll. 9- XII-2014, Panhala, A.S. Desai; legs on slides, labeled as above.

Paratype- 1 Females, Sex ratio (M: F) 0:1, July 2011- February 2015; legs on slides, rest part pinned, preserved in insect box.

Etymology

The species name is due to the scutellar characters.

Distributional Record

India, Maharashtra, ♀ 1 Panhala 9- XII-2014.

Remarks

According to key of Crosskey (1976) and Gilasian *et al.* (2013) this species runs close to *Cylindromyia rubida* (Loew, 1854) by having following characters:

1. Antenna as long as face.
2. Frontal vita, legs and basicosta yellow in color.
3. Katepisternum with 2 setae.
4. Abdominal tergites without discal setae.
5. Outer vertical setae absent.
6. Proepisternum bare.

However, it differs from the above species by following distinct characters:

1. Scutellum brownish in color.
2. Scutellum shows longer subapical setae.
3. Mid tibia with 1 anterodorsal setae.
4. Hind tibia with 1 posteroventral setae.
5. Tergite 3rd, 4th and 5th shows dusting.
6. Antennal formula: S L/W=0.818, P L/W=3.153, F L/W=10.8, A=4.923.

***Phryxe carbonaria* sp.n.**

(Plate 22, Figures 181 to 189)

Male (Figures 181-189)

Body 9.12 mm in length, 4.11 mm in width; wing 6.73mm in length, 2.45mm in width; antenna 1.74 mm long; halter 1.03mm long, 0.35mm wide.

Head (Figure 183)

1.65mm in length, 2.49mm in width; vertex wide; inner vertical bristles present; outer vertical bristle absent; ocelli present; ocellar bristles 2 setae proclinate manner; frontal vita well developed; frontal bristles medioclinate; fronto orbital plate with one row of medioclinate strong setae with few hairs; reclinate upper orbital bristles with 2 pairs; proclinate orbital bristles absent; face concave; lower facial margin less concave, protruted; vibrissa present, strong and small; facial ridge bare; parafacial with small hairs; gena wider than profrons; genal dilation scarely developed; back of head concave, pale yellow hairs. **Eyes** 0.95mm in length, 2.12mm in width reddish brown, haired. **Antenna** (Figure) scape short, 0. 11mm in length,0.10mm in width, pedicel 0.29mm in length, 0.12mm in width; first flagellomere 1.34mm in length, 0.14mm in width. **Arista** long, straight and bare; first and second aristomeres absent. **Mouth** proboscis well developed; prementum short with labella; palpus filiform, parallel sided.

Antennal Formula

S L/W=1.1, P L/W=2.416, F L/W=9.571, A=4.362.

Thorax (Figure 185)

3.40mm in length, 2.79mm in width; humeral callus 3 reclinate setae; proepisternum with hairs; proepimeron 2 setae (1 prostagmitic setae and 1 propleural setae present); prosternum with hairs. **Scutum** 2.49mm in length, 2.72mm in width; acrostichal bristles long, strong and reclinate (presutural 3+ postsutural 3); dorsal central bristles long, strong and reclinate (presutural 3+ postsutural 4); notopleuron 2 setae reclinate; postalar callus 2 setae. **Scutellum** 0.91mm in length, 1.77 m in width; completely black; basal scutellar bristles present very strong and slightly curved inwardly; lateral scutellar bristle small weak than other; subapical scutellar bristles divergent, very long and strong; apical scutellar bristles horizontal; discal scutellar bristles short. **Subscutellum** convex; katepisternum 4 sternopleuron setae; anepimeron 1 setae; katepimeron bare; meron with row of setae; postmetacoxal area absent; anatergite bare; anterior thoracic spiracle narrow and closed by fringes of hairs.

Wing (Figure 187)

6.73mm in length, 2.45mm in width, transparent with brownish dark pattern; lower calypter well developed its inner margin contagious to lateral margin of scutellum, margin seen dorsally; second costal section bare ventrally; base of costa without costigial bristles; r_1 bare; cu a_1 bare; bend of m distinct; wing cell r_{4+5} open; cross vein r-m present; cross vein d m-cu present; anal vein present but not reaching

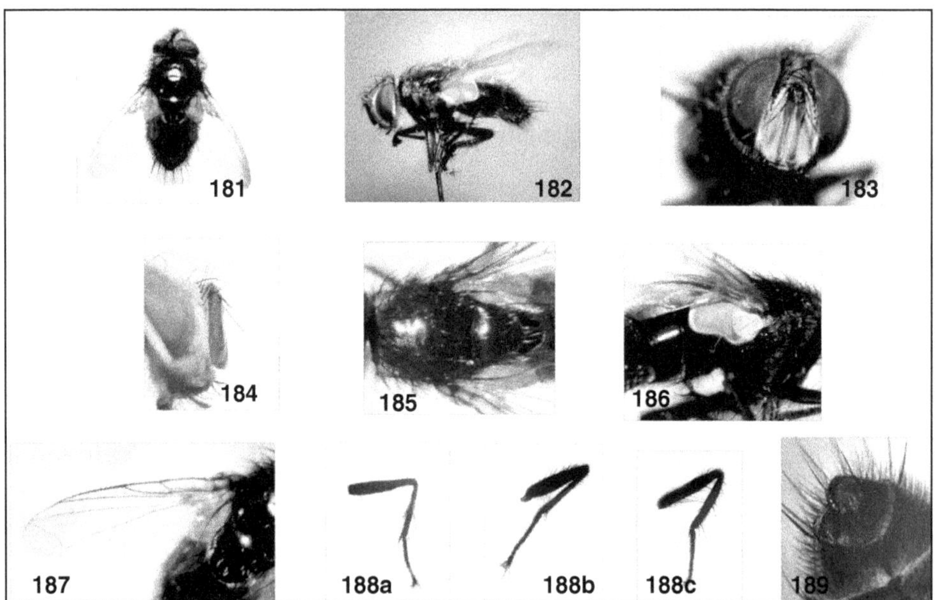

Plate 22. Figure 181: *Phryxe carbonaria* sp.n. (Male) Dorsal view; Figure 182: Lateral View; Figure 183: Frontal View of Head; Figure 184: Antennae; Figure 185: Dorsal View of Thorax; Figure 186: Lateral View of Thorax; Figure 187: Fore Wing; Figure 188a: Fore Leg; Figure 188b: Mid Leg; Figure 188c: Hind Leg; Figure 189: Genitalia.

towards hind margin of wing. **Halter** reddish yellow, 1.03mm in length, 0.35mm in width.

Leg

Fore and mid femora with two rows of setae ventrally. **Fore leg** (Figure 188a) femora 2.34mm; tibia 2.54mm and tarsus 1.50mm; preapical anteriodorsal setae present. **Mid leg** (Figure 188b) femora 2.25mm; tibia 2.44mm and tarsus 1.42mm; **mid** tibia pre apical anterior dorsal very long spiny, pre apical posterior dorsal very long spiny. **Hind leg** (Figure 188c) femora 2.29mm; tibia 2.47mm and tarsus 1.39mm; hind tibia, preapical posterioventral seta present small, preapical anteriodorsal seta long spiny; antero dorsal bristles comb like, with spiny setae; hind coxa with 1 seta.

Abdomen

4.07mm in length, 3.19mm in width; abdomen entirely shining black; abdominal tergite 5th as long as 4th tergite; mid dorsal depression syntergite 1+ 2 up to 2nd tergite margin; discal bristles absent on tergite 3rd and 4th; abdominal sternites overlapped ventrally by last segment.

Genitalia (Figure 189)

Male terminalia retracted within abdominal tergite 5; tergite 6th less reduced joined with 7th tergite. Cercus and syrstylus present below syntergite 7+8.

Colour

Black: Head, thorax abdomen, fore tibia, mid tibia, hind tibia, palpus.

Brown: Frontal vita, antenna.

Silvery white: Face, parafacial.

Reddish brown: Eyes.

Host- Unknown.

Host Plant- unknown.

Paratype- 1 Males, Sex ratio (M: F) 1:0, July 2011- February 2015; legs on slides, rest part pinned, preserved in insect box.

Etymology- The species name *'carbonaria'* is due to the black color to body of species.

Distributional Record

India, Maharashtra, ♂ 1 Radhanagari 22- XI-2013.

Remarks

According to key of Crosskey (1976) and Tschorsnig and Herting (2005) this species closely runs with *Phryxe prima* (Brauer and Bergenstamm, 1889).

1. Four dorsocentral setae.
2. Cheeks downwards not much narrowed, their narrow point as wide as 0.50 to 0.67x of 3rd antennal segment.

However, it differs from following characters:

1. Arista thickened at least 1/3 to 1/4th in males of its length.
2. Ocellar setae very strong.
3. Third antennal segment slightly concave near to the base of arista.
4. Base of 3rd antennal segment reaches to epistomal margin.
5. Presence of 2 pair of lateral scutellar bristles.
6. **Antennal formula:** S L/W=1.1, P L/W=2.416, F L/W=9.571, A=4.362.

Carcelia (Carcelia) abdominalis sp.n.

(Plate 24, Figures 190 to 198)

Male (Figures 190-198)

Body 09.82 mm long, 1.09 mm broad, wing 7.27mm in length, 2.46 mm in width, antenna 1.33 mm long, Halter 1.19mm length, 0.47mm width.

Head (Figures 192 and 193)

1.87mm in length, 2.32mm in width; vertex wide; inner vertical bristles single pair reclinate; outer vertical bristle present short reclinate; ocelli present; ocellar bristles strong, proclinate; frontal vita well developed; frontal bristles small, medioclinate; fronto orbital plate well developed, with 2 pairs of proclinate setae; two pairs of strong lateroclinate upper orbital bristles; proclinate orbital bristles two pairs; face concave, white; lower facial margin slightly visible in lateral view; vibrissa strong; facial ridge straight, not seen in lateral view; parafacial (cheeks) below frontal bristles bare, very narrow; gena not visible in the lateral view; peristome narrower; genal dilation not clearly visible; back of head with pale hairs. **Eyes** (Figure)0.98 mm in length, 2.63mm in width haired, reddish brown; **Antenna** (Figure)scape 0. 09mm in length, 0.11mm in width, pedicel short, 0. 21mm in length, 0.15mm in width; first flagellomere 1.03mm in length, 0.15mm in width. **Arista** present near base of first flagellomere, cylindrically thickened. **Mouth** proboscis well developed; prementum with short labella; palpus well developed. **Antennal formula:** S L/W=0.819, P L/W=1.40, F L/W=6.416, A=3.334.

Thorax (Figures 194 and 195)

3.46mm in length, 2.29mm in width; humeral callus with 3 setae; proepisternum bare; proepimeron with 2 setae. **Scutum** 2.44 mm in length, 2.31mm in width with 4 strips; acrostichal bristles (presutural 2 + postsutural 3); dorsal central bristles (presutural 2 + postsutural 4); notopleuron with 2 reclinate setae; postalar callus with 2 setae. **Scutellum** 1.12 mm in length, 1.34mm in width; basal scutellar bristles very strong and slightly curved inwardly; lateral scutellar bristle small, weaker than basal setae; subapical scutellar very long, wide apart, strong, divergent; apical scutellar bristles very small, weak, crossed. **Subscutellum** convex; katepisternum with 2 sternopleuron setae; anepimeron with single setae; katepimeron bare; meron with row of setae; anatergite bare; anterior thoracic spiracle narrow closed with fringes hairs.

Wing (Figure 196)

7.27mm in length, 2.46 mm in width, transparent; lower calypter well developed its inner margin less contiguous to lateral margin of scutellum; second costal section bare; base of costa with hairs, slightly lightened; forth costal section longer than 6th section; r_1 bare; cu a_1 bare; bend of m distinct; wing cell r_{4+5} open, without hairs at base; cross vein r-m present; cross vein d m-cu present; anal vein present but not reaching towards hind margin of wing. **Halter** reddish yellow, 1.19mm length, 0.47mm width.

Leg (Figures 197a-c)

Fore and mid femora with row of spines ventrally; **Forelegs** (Figure 197a) femora 3.02mm; tibia 2.83mm and tarsus 3.04mm; one preapical anteriodorsal setae present; fore tibia without bristle. **Mid legs** (Figure 197b) Femora 2.89mm; tibia 2.75mm, tarsus 3mm; mid tibia with 2 anterodorsal setae; preapical anteriodorsal setae. **Hind legs** (Figure 197c)femora 3.03mm; tibia 2.71mm, tarsus 3.09mm; hind tibia with pair of strong anteriodorsal setae; hind coxa setulose.

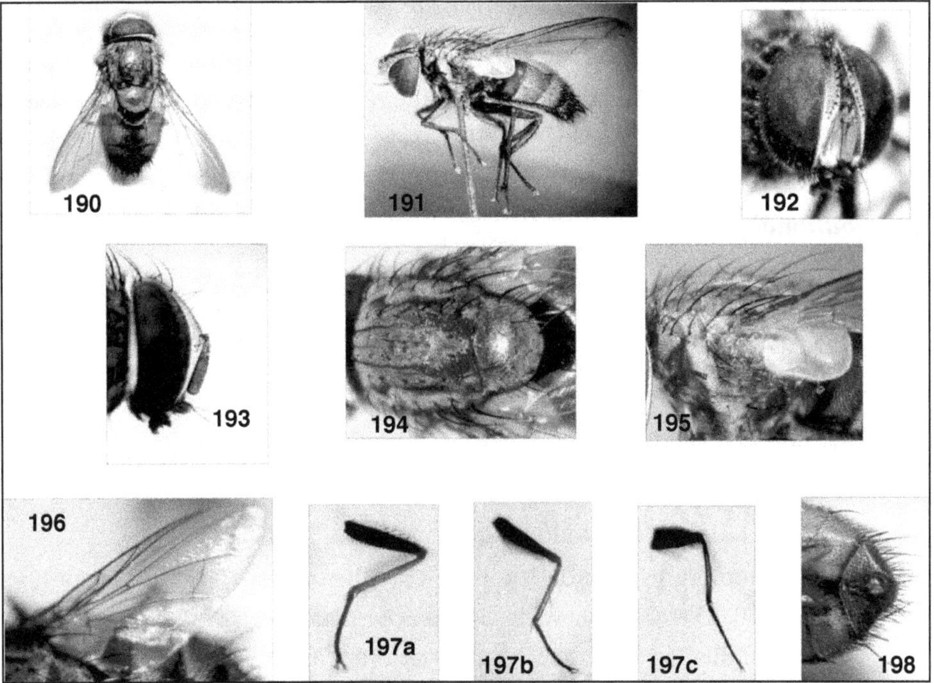

Plate 24. Figure 190: *Carcelia abdominalis* sp.n. (Male) Dorsal view; Figure 191: Lateral View; Figure 192: Frontal View of Head; Figure 193: Antennae; Figure 194: Dorsal View of Thorax; Figure 195: Lateral View of Thorax; Figure 196: Fore Wing; Figure 197a: Fore Leg; Figure 197b: Mid Leg; Figure 197c: Hind Leg; Figure 198: Genitalia.

Abdomen (Figures 190-191)

4.49mm in length, 3.07 mm in width, oval; abdominal tergite 5th as long as 4th abdominal tergite; mid dorsal depression reaching towards 1+ 2 syntergite; median discal bristles absent; abdominal sternites overlapped by ventral edges of tergites; 5th tergite dusted.

Genitalia (Figure 198)

Terminalia retracted within abdominal tergite 5th; tergite 6th reduced attaches to terigite 7th and 8th; surstylus and cercus present below epandrium.

Colour

Black: Thorax,

Dark brown: Frontal vita, antenna

Yellow: Abdomen, palpus, fore tibia, mid tibia, hind tibia

Silvery white: Head, face, parafacial

Reddish brown: Eyes

Host- Unknown

Host Plant- *Nepeta indica*

Paratype- 3 Males, Sex ratio (M: F) 3:0, July, 2011- February, 2015; legs on slides, rest part pinned, preserved in insect box.

Etymology- The species name is due to abdomenal features of host plant.

Distributional Record

India, Maharashtra, ♂ 1 Panhala 15-VII- 2013; ♂ 1 Radhanagari 10-XI- 2013

Remarks

According to key of Crosskey, 1976 and Tschorsnig and Herting, 2005, this species closely runs with *Carcelia bombylans* R. and D.

1. Humeral callus partially yellowish brown
2. Dusting yellowish orange

However, it differs from following characters:

1. Hairs of tergite 3rd and 4th with 1/4th of the corresponding segment
2. Frons 0. 25 to 0.35x as wide as one eye in males
3. Partial dusting seen on 3rd and 4th tergite
4. Marginal setae of tergite t_{1+2} absent
5. **Antennal formula:** S L/W=0.819, P L/W=1.40, F L/W=6.416, A=3.334

Sub-family-Exoristinae

Tribe Exoristini

Exorista larvarum (Linnaeus, 1758)

(Plate 25 Figures 199 to 207)

Female (Figures 199-207)

Body 13.51mm in length, 3.44 mm in width; wing 9.10mm in length, 4.01mm in width; antenna 1.53 mm long; halter 1.61mm in length, 0.59mm in width.

Head (Figures 201 and 202)

2.15 mm in length, 2.98mm in width; vertex wide; one pair inner vertical bristles very strong, reclinate; outer vertical bristle straight; ocelli present; ocellar bristles 2 long, proclinate as; frontal vita well developed; frontal bristles medioclinate; fronto orbital plate with single row of medioclinate setae; lateroclinate with pair of strong reclinate setaelong as inner vertical setae; proclinate orbital bristles absent; parafrontalia yellowish golden; lower facial margin less visible in lateral view; vibrissa paired strong, setulae above the vibrissa do not rise to the lower frontal bristle; facial ridge with strong setal row at margin; parafacial with small hairs; gena present narrow; genal dilation well developed; back of head dense pale white hairs at back. **Eyes** 2.45mm in length, 1.36mm in width, bare, reddish-brown. **Antenna**) scape very short 0.11mm in length, 0.09mm in width; pedicel short 0.25mm in length, 0.12 mm in width; short with small hairs; first flagellomere 1.18mm in length, 0.14mm in width, concave with pointed tip; **arista** long tapering (apex pointed + base little broad). **Mouth** Proboscis well developed; prementum short; palpus slightly swollen apically.

Antennal Formula

S L/W=1.222, P L/W=2.083, F L/W=8.428, A= 3.911.

Thorax (Figures 203 and 204)

5.04mm in length, 3.40mm in width; humeral callus 3 reclinate setae; prosternum setulose; proepisternum bare; proepimeron with 2 setae. **Scutum** 3.51mm in length, 3.46mm in width; acrostichal bristles (presutural 3+ postsutural 4); dorsal central bristles (presutural 3+ postsutural 4); 4 (1+3) intra alar setae; notopleuron 2strong setae; postalar callus 2 setae. **Scutellum** 1.53mmin length, 2.22mm in width, reddish; basal scutellar bristles present; lateral scutellar bristle present; subapical scutellar bristles long, parallel; apical scutellar bristles small; discal scutellar bristles small; subscutellum convex; katepisternum with 2 setae and small hairs; anepimeron with single strong setae; katepimeron bare; meron small with row of setae; postmetacoxal area present; anatergite bare; anterior thoracic spiracle narrow and closed by fringes of hairs.

Wing (Figure 205)

9.10mm in length, 4.01mm in width; lower calypter well developed its inner margin less contagious to lateral margin of scutellum; second costal section bare; base of costa with small spine (setulae); costal bristles present; r_1 bare; cu a_1 bare; bend of m distinct, curved at base; wing cell r_{4+5} open with few hairs at base; cross vein r-m present; cross vein d m-cu present; anal vein present but not reaching towards hind margin of wing; halter reddish brown, 1.61mm length, 0.59mm width.

Leg (Figures 206a-c)

Fore and mid femora with two rows of spines ventrally, silvery white; **Foreleg** (Figure 206a) femora 3.38mm; tibia 2.51mm, tarsus 2.81mm; fore tibia with 1 preapical anteriodorsal setae present,1 preapical posteriodorsal setae present, 1 long posterioventral setae. **Mid leg** (Figure 206b) femora 3.31mm; tibia 2.39mm, tarsus 2.78mm; mid tibia with 3 anterior dorsal setae, 2 posteriodorsal setae. **Hind leg** (Figure 206c) femora 3.35mm; tibia 2.43mm and tarsus 2.77mm; hind tibia with 1 preapical anterioventral setae, 1 preapical anteriodorsal seta small, and 3 long antero dorsal bristles; hind coxa with few setae.

Abdomen (Figures 199-200)

6.32mm in length, 3.93mm in width, with black central longitudinal stripe visible in lateral view; Abdominal tergite 5th as long as 4th tergite, also tergite 5th; Mid

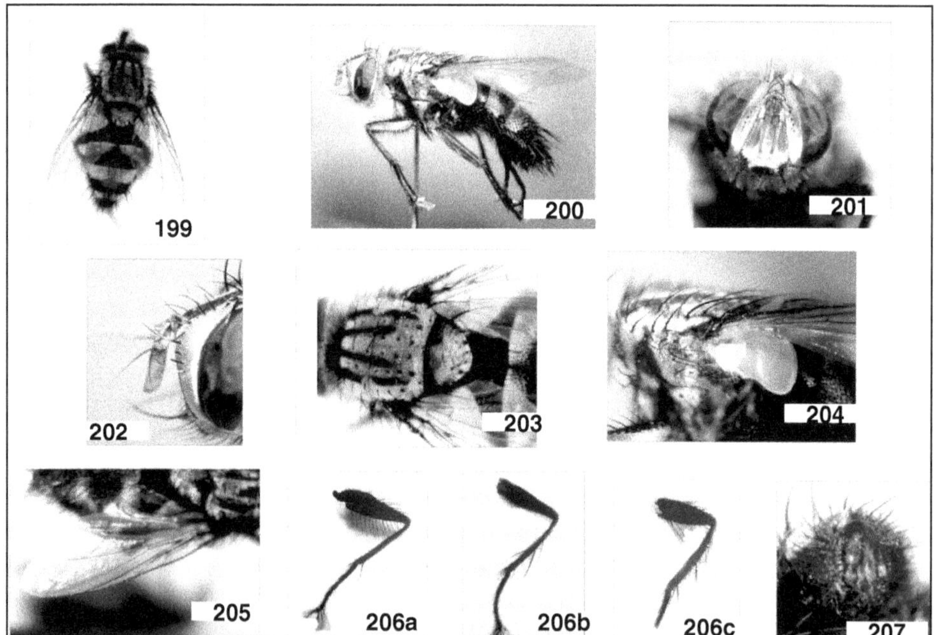

Plate 25. Figure 199: *Exorista larvarum* Linnaeus (Female) Dorsal view; Figure 200: Lateral View; Figure 201: Frontal View of Head; Figure 202: Antennae; Figure 203: Dorsal View of Thorax; Figure 204: Lateral View of Thorax; Figure 205: Fore Wing; Figure 206a: Fore Leg; Figure 206b: Mid Leg; Figure 206c: Hind Leg; Figure 207: Genitalia.

dorsal depression 1+2 syntergite excavate to hind margin; Discal bristles present; Abdominal sternites overlapped by ventral edges of tergites.

Genitalia (Figure 207)

Terminalia retracted within abdominal tergite 5th; Tergite 6 less reduced joining syntegite 7th +8th. Piercer covered with epandrium.

Colour

Black: Thorax, Abdomen, Mid Tibia, Hind Tibia.

Brown: Antenna, Palpus.

Dark brown: Frontal vita.

Silvery white: Head, Face, Parafacial, Fore Tibia.

Reddish brown: Eyes.

Host- Unknown.

Host Plant- Unknown.

Paratype- 4 Females, Sex ratio (M: F) 0: 4, July 2011- February 2015; legs on slides, rest part pinned, preserved in insect box.

Distributional Record

India, Maharashtra, ♀1 Karveer 08-IX-2013; ♀ 1 Vita 10-IX- 2013; ♀ 1 Phaltan 05-IX- 2014; ♀ 2 Vengurla 21-XI- 2014.

Remarks

According to key of Crosskey (1976) and Tschorsnig and Herting (2005) this species is *Exorista larvarum* (Linnaeus, 1758).

Sub-family-Exoristinae

Tribe-Blondelini

Compsilura concinnata (Meigen, 1824)

(Plate 29, Figure 208 to 216)

Male (Figures 208-216)

Body 9.49 mm in length, 3.90mm in width; wing 5.82mm in length, 3.10mm in width; antenna 1.57 mm in length; Halter 1.23mm in length, 0.36mm in width.

Head (Figures 210 and 211)

1.75 mm in length, 2.12mm in width; vertex wide; inner vertical bristles with one reclinate setae; outer vertical bristle present; ocelli present; ocellar bristles present; frontal vitta very narrow; frontal bristles medioclinate; fronto orbital plate with row medioclinate setae; reclinate upper orbital bristles with three strong setae; proclinate orbital bristles with two strong setae; face concave; lower facial margin not visible in lateral view; vibrissa strong; facial ridge concave; parafacial bare; gena wide;

genal dilation well developed; back of head slightly concave, with white hairs. **Eyes** 1.45mm in length, 0.92 mm in width, hairy, brownish. **Antenna** (Figure 211) scape 0.09mmin length, 0.08mm in width; pedicel 0.12mm in length, 0.11mm in width; first flagellomere 1.36mm in length, 0.14mm in width, concave. **Arista** inserted at base of first flagellomere. **Mouth** proboscis well developed; prementum with short labella; palpus well developed slightly swollen at apex.

Antennal Formula

S L/W=1.125, P L/W=1.091, F L/W=9.714, A=3.976.

Thorax (Figures 212 and 213)

3.76mm in length, 3.43mm in width; humeral callus with 3 reclinate setae; proepisternum (propleuron) hairy; proepimeron with 3 setae dorsally projected; prosternum hairy. **Scutum** 2.64mm in length, 2.55mm in width, 4 dark longitudinal strips; acrostichal bristles reclinate (presutural 3 + postsutural 3); dorsal central bristles reclinate setae (presutural 3+ postsutural 3), with few setulae; notopleuron with 2 reclinate setae; postalar callus with 2 reclinate setae. **Scutellum** 1.12mm in length, 0.88mm in width; basal scutellar bristles slightly curved, short, lateral scutellar bristles as long as half of subapical scutellar bristles; subapical scutellar bristles divergent, longer; apical scutellar bristles slightly crossed. **Subscutellum** convex; katepisternum with 3 strong setae (1 reclinate + 1 proclinate); anepimeron

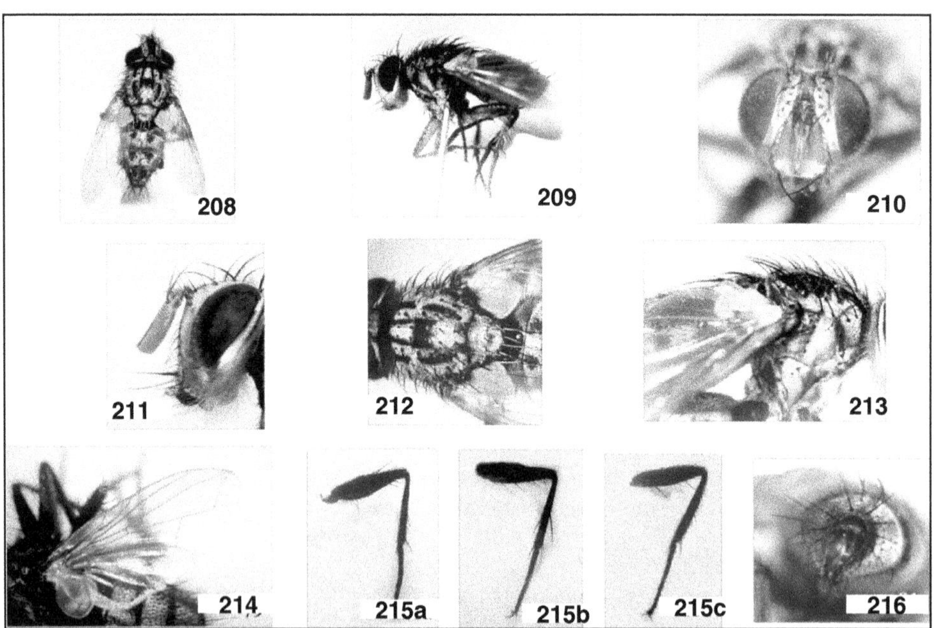

Plate 26. Figure 208: *Compsilura concinnata* Meigen (Male) Dorsal view; Figure 209: Lateral View; Figure 210: Frontal View of Head; Figure 211: Antennae; Figure 212: Dorsal View of Thorax; Figure 213: Lateral View of Thorax; Figure 214: Fore Wing; Figure 215a: Fore Leg; Figure 215b: Mid Leg; Figure 215c: Hind Leg; Figure 216: Genitalia.

with 1 setae; katepimeron bare; meron with row of hairs; anatergite bare; anterior thoracic spiracle narrow and closed by fringes of hairs.

Wing (Figure 214)

5.82mm in length, 3.10mm in width, transparent; lower calypter well developed inner margin less contiguous to lateral margin of scutellum; second costal section bare; base of costa with small setulae; costal bristles absent; fourth coastal section longer than 6th costal section; r_1 bare; cu a_1 bare; bend of with m obtuse angle; wing cell r_{4+5} open at wing margin; cross vein r-m present; cross vein d m-cu present; anal vein not reaching hind margin of wing. **Halter** reddish yellow, 1.23mm length, 0.36mm width.

Leg (Figures 215a-c)

Fore and mid femora with ventral spiny rows of bristles. **Forelegs** (Figure 215a) femora 1.96mm; tibia 2.03mm, tarsus 1.56mm fore tibia with one preapical anteriodorsal setae. **Midlegs** (Figure 215b) femora 1.94mm; tibia 1.99mm, tarsus 1.53mm mid tibia with one strong anteriodorsal setae; one strong anterioventral setae. **Hind legs** (Figure 215c) femora 1.97mm; tibia 1.96mm, tarsus 1.55mm, hind tibia with 1 anterodorsal setae; one preapical anteriodorsal setae; hind coxa with 1 small seta.

Abdomen (Figures 208 and 209)

3.98mm in length, 2.77mm in width; mid dorsal depression reaches back to hind margin of that segment; discal bristles present in 3rd and 4th tergite, abdominal sternite overlapped by the ventral edges of tergite.

Genitalia (Figure 216)

Male terminalia present below abdominal tergite 5th; tergite 6th more reduced joining segment 7th. tergite 7th and 8th fused and forms single segment (syntergite 7th + 8th); cercus and surstylus are present below tergite 7th and 8th.

Colour

Brown: Palpus, thorax, abdomen.

Blakish-brown: Fore tibia, mid tibia, hind tibia.

Silvery white: Head, face, parafacial, frontal vita, antenna.

Reddish brown: Eyes.

Host- Unknown.

Host Plant-Unknown.

Paratype- 6 Males, Sex ratio (M: F) 6:0, July 2011- February 2015; legs on slides, rest part pinned, preserved in insect box.

Distributional Record

India, Maharashtra, ♂ 2 Radhanagari 16- VII-2013; ♂ 1 Shahuwadi 08-XI-2013; ♂ 1 Gaganbawada 9-XI- 2014; ♂ 1 Gargoti 22- VI-2015; ♂ 1 Koyna 01- VII-2015.

Remarks

According to key of Crosskey, 1976 this species is *Compsilura concinnata* (Meigen, 1824).

Sub-family-Exoristinae

Tribe Goniinae

Leschenaultia ciliata (Macquart, 1848)

(Plate 27, Figures 217 to 225)

Male (Figures 217-225)

Body 14. 96 mm in length, 6.6mm in width; wing 10.12mm in length, 4.11mm in width; antenna 1.74 mm long; halter 1.38mm in length, 0.44mm in width.

Head (Figures 219 and 220)

2.14mm in length, 3.66mm in width; vertex very wide; single pair of strong, reclinate inner vertical bristles; outer vertical bristle single pair slightly reclinate; ocelli present; ocellar bristles present; frontal vitta present; frontal bristles medioclinate, strong, small; fronto orbital plate well developed, with medioclinate row of setae also with small hairs scattered on it; lateroclinate upper orbital bristles with single pair; proclinate orbital bristles single pair; face concave; lower facial margin present; vibrissa short, crossed; facial ridge slightly concave; parafacial entirely bare; gena present; genal dilation present; back of head with pale hairs, dense. **Eyes** 1.08 mm in length, 3.37mm in width, brownish; **Antenna** (Figure 220) scape short 0.12mm in length, 0.11mm in width, pedicel 0.32mm in length,0.15mm in width; first flagellomere concave, 1.03mm in length, 0.14mm in width. **Arista** arises at the base of flagellomere, tapering; **Mouth** proboscis seldom, very small; prementum short; palpus slightly swollen at apex.

Antennal Formula

S L/W=1.09, P L/W=2.133, F L/W=7.357, A=3.526.

Thorax (Figures 221 and 222)

Black, 5.65mm in length, 4.59mm in width; humeral callus five setae; proepisternum bare; proepimeron with two setae. **Scutum** 3.75 mm in length,4.49 mm in width, with 4 dark strips; acrostichal bristles small weak (presutural 2 + postsutural 3); dorsal central bristles small weak (presutural 2 + postsutural 3); notopleuron with 2 setae; postalar callus with 2 setae. **Scutellum** 1.90 mm in length, 3.17mm in width; basal scutellar bristles slightly bend; lateral scutellar bristle lateral strong, subapical scutellar bristles long, strong slightly divergent; apical scutellar bristles crossed paired, strong. **Subscutellum** convex; katepisternum with 3 setae; anepimeron with strong setulae or hairs; katepimeron bare; meron with row of setae; postmetacoxal area present; anatergite bare; anterior thoracic spiracle narrow closed with fringes hairs.

Wing (Figure 223)

10.12mm in length, 4.11mm in width, transparent; lower calypter well developed its inner margin less contiguous to lateral margin of scutellum; second costal section present bare; base of costa with few small hairs; forth costal section bare longer than 6th section; r_1 bare; cu a_1 bare; bend of m distinct; wing cell r_{4+5} open, single setulae at base; cross vein r-m present; cross vein d m-cu present; anal vein present but not reaching towards hind margin of wing. **Halter** reddish yellow, 1.38mm length, 0.44mm width.

Leg

Fore leg (Figure 224a) femora 3.21mm; tibia 3.27mm, tarsus 3.25mm; fore tibia with one pair of preapical anteriodorsal setae. **Mid leg** (Figure 224b) femora 3.24mm; tibia 3.19mm, tarsus 3.22mm; mid tibia with preapical anteriodorsal setae. **Hind leg** (Figure 224c) femora 3.17mm; tibia 3.20mm, tarsus 3.19mm; anteriodorsal setae of hind tibia nearly touching to each other, anteriodorsal bristles comb like dorsally; hind coxa with 2 small setae.

Abdomen (Figures 217 and 218)

6.90mm in length, 5.14mm in width, oval, black; abdominal tergite 5th shorter than 4th tergite; tergite 4th with patch of dense hairs; mid dorsal depression crossed

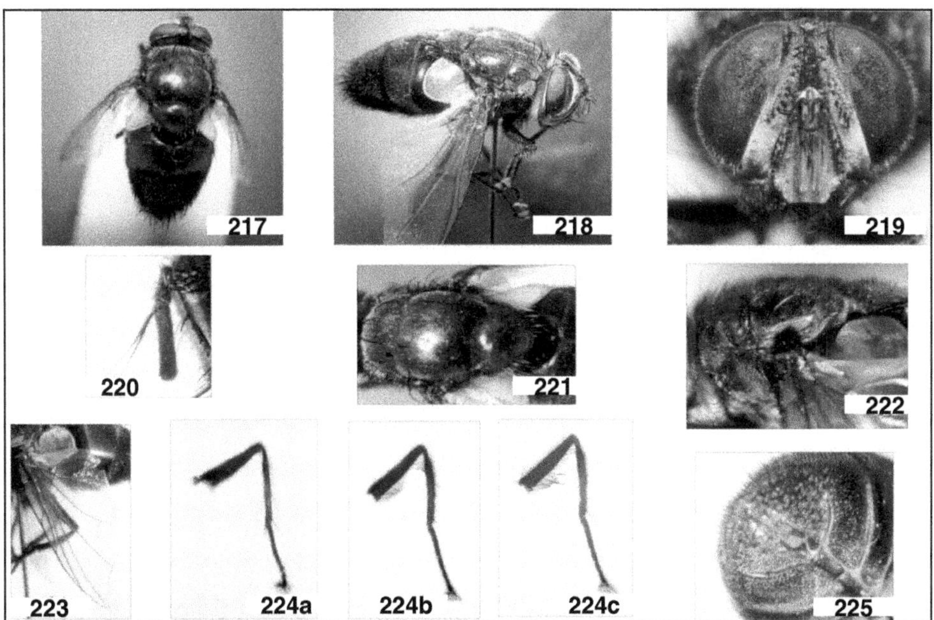

Plate 27. Figure 217: *Leschenaultia ciliata* Macquart (Male) Dorsal view; Figure 218: Lateral View; Figure 219: Frontal View of Head; Figure 220: Antennae; Figure 221: Dorsal View of Thorax; Figure 212: Lateral View of Thorax; Figure 223: Fore Wing; Figure 224a: Fore Leg; Figure 224b: Mid Leg; Figure 224c: Hind Leg; Figure 225: Genitalia.

1+2 syntergite margin on second tergite, 4 pairs of marginal bristles; discal bristles absent; abdominal sternites overlapped by ventral edges of tergites.

Genitalia (Figure 225)

Terminalia retracted within abdominal tergite 5th; Tergite 6th less reduced, joining segment 7th; cercus and surstylus are hidden below epandrium.

Colour

Black: Thorax, abdomen, fore tibia, mid tibia, hind tibia.

Brown: Antenna, palpus.

Dark brown: Frontal vita.

Silvery white: Head, Face, parafacial.

Reddish brown: Eyes.

Host- Unknown.

Host Plant- Unknown.

Paratype- 1 Males, Sex ratio (M: F) 1:0, July 2011- February 2015; legs on slide, rest part pinned, preserved in insect box.

Distributional Record

India, Maharashtra; ♂1 Gaganbwada 08-XI-2013.

Remarks

According to key of Crosskey (1976) and Tschorsnig and Herting (2005) this species is *Leschenaultia ciliata* (Macquart, 1848).

5

Abundance and Distribution of Tachinids

Introduction

Tachinid flies are excellent biocontrol agents recorded from agro and forest ecosystems. Hence, they were studied with respect to distribution and seasonal abundance. Seasonal abundance and species richness implies exact time of occurrence and probable number of species existing in specific area. Biological control of insect pest has attracted the attention of farmers, agriculturist, researchers and scientists since it is ecofriendly. Biological control has ecological basis hence, seasonal abundance and distribution of tachinids will add great relevance in ecofriendly pest control.

Perusal of literature indicates that very little attention is paid on the study of seasonal abundance and distributional record of tachinid flies from India including Western Maharashtra. Crosskey (1976), Das (1993), Lahiri (2003, 2006), Sathe (2012; 2014), Sathe *et al.* (2014), Shendge and Sathe (2014), etc have attempted the study of seasonal abundance and distributional records of tachinid flies.

Materials and Methods

Seasonal abundance and distributional records of tachinid flies have been studied by spot observations from different places of Kolhapur, Sangli, Satara and Sindhudurg districts (Figure 17) at 15 days intervals during the years of 2011-2015 by adopting one man one hour search method. The methodology is given in the Chapter 3, Materials and Methods.

Results

Results recorded in Tables 1–4 shows that, 20 species of tachinid flies from four subfamilies *i.e.* Tachininae, Exoristinae, Dexiinae and Phasiinae were reported from Kolhapur, Sangli, Satara and Sindhudurg districts.

Table 1 shows seasonal abundance of tachinid flies in Kolhapur district. Out of 20 species, 19 species were reported from Kolhapur district. *T. nigripes, D. vacua, D. indica* and *P. siberita* shows maximum population. *C. scutellata* and *Carcelia abdominalis* was rare species. From Kolhapur district Panhala, Radhanagari and Gaganbawada are near to Western Ghats which were good spot for tachinid population. Minimum diversity of tachinids was reported from Karveer and Kagal. Hatakanangale showed less diversity of tachinid flies because of dried area. Average population was reported from Shahuwadi, Chandgad, Kagal, Bhogawati and Gargoti regions. Population peak was observed maximum during August, September and November. Low population was observed during months of January to May because of summer and less availability of host species. The dominant sub-family in Kolhapur district was Exoristinae (Figure 226).

Table 2 shows distributional pattern and seasonal abundance of tachinid flies in Sangli district. Out of 20 species, Only 08 species were reported from Sangli district (Table 2). *T. nigripes* showed maximum population. Rest of all species was less in population. Sangli district showed comparatively less tachinid population to the other districts. At study spots Jath, Sonwade and Walawa tachinids were rarely available. Population of tachinids was abundant during August, October and November was low during months of January to June due to very poor rainfall and less availability of host species. The dominant sub-family in sangli district was Exoristinae (Figure 226).

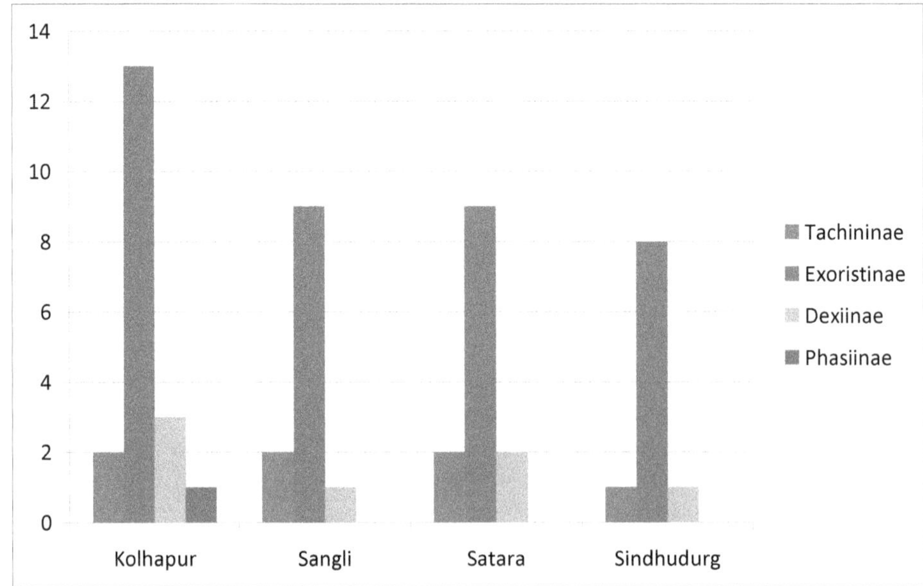

Figure 226: Occurrence of Tachinid Subfamilies from Selected Districts of Western Maharashtra.

Table 1: Seasonal Abundance of Tachinid Flies from Kolhapur District (2011-2015)

Sl.No.	*Name of Tachinid Fly*	*Jan.*	*Feb.*	*March*	*April*	*May*	*June*	*July*	*Aug.*	*Sept.*	*Oct.*	*Nov.*	*Dec.*
	Sub-family-Tachininae												
	Tribe- Tachinini												
1.	*Peleteria tricolor* sp. n.	–	–	–	–	–	–	–	+ Δ	–	+	–	–
	Tribe- Thelairini												
2.	*Thelaira nigripes* F.	–	–	–	–	–	–	–	–	–	–	–	–
	Tribe- Phyllomyini												
3.	*Phyllomya panhalaensis* sp.n.	–	–	–	–	–	–	–	+	–	–	+	–
	Sub-family-Exoristinae												
	Tribe-Bloneliini												
4.	*Urodexia uramyoides* Town.	–	–	–	–	–	–	+	–	+	–	–	–
5.	*Compsilura concinnata* (Meigen, 1824).	–	–	–	–	–	–	+	+	+	–	–	–
	Tribe- Exoristini												
6.	*Ctenophorinia dorsalis* sp. n.	–	–	–	–	–	–	–	–	+	–	–	–
7.	*Exorista larvarum*	–	–	–	–	–	–	–	–	+	–	+	–
	Tribe- Winthemiini												
8.	*Winthemia chandgadensis* sp. n.	–	–	–	–	–	–	–	–	–	–	+	–
	Tribe -Carceliini												
9.	*Carcelia (Carcelia) orientalis* sp.n.	–	–	–	–	–	–	+ Δ	+	–	–	–	–
10.	*Carcelia (Carcelia) sataraensis* sp. n	–	–	–	–	–	–	–	+	–	–	–	–
11.	*Carcelia (Carcelia) abdominalis* sp.n.	–	–	–	–	–	–	+	+	–	–	–	–
12.	*Carcelia* (*Senometopia*) *kolhapurensis* sp. n.	–	–	–	–	–	–	–	–	+	–	+	–

Contd...

Table 1–*Contd...*

Sl.No.	*Name of Tachinid Fly*	*Jan.*	*Feb.*	*March*	*April*	*May*	*June*	*July*	*Aug.*	*Sept.*	*Oct.*	*Nov.*	*Dec.*
	Tribe- Goniini												
13.	*Leschenaultia ciliata* (Macquart, 1848)	–	–	–	–	–	–	–	–	–	–	+	–
	Tribe- Sturmini												
14.	*Blepharipa schineri* Mensil	–	–	–	–	–	–	–	+	–	–	–	–
	Tribe- Eryciini												
15.	*Anaeogmena striatalis* sp. n	–	–	–	–	–	–	–	–	+ Δ	–	+	–
16.	*Phryxe carbonaria* sp.n.	–	–	–	–	–	–	–	+	–	–	–	–
	Sub-family- Dexiinae												
	Tribe- Dexiini												
17.	*Prosena siberita* F.	–	–	–	–	–	–	+ Δ	–	–	–	–	–
	Tribe- Dexia												
18.	*Dexia vacua* Fall.	–	–	–	–	–	–	+	+ Δ			+	–
19.	*Dexia indica* sp,n	–	–	–	–	–	–	+ Δ	–	–	+	–	–
	Sub-family- Phasiinae												
	Tribe- Cylindromyini												
20.	*Cylindromyia scutellata* sp.n. *	–	–	–	–	–	–	–	–	–	–	–	+

+: Indicates that Species present; –: Indicates that species absent, Δ indicates that maximum population, *: Indicates that rare species.

Table 2: Seasonal Abundance of Tachinid Flies from Sangli District (2011-2015)

Sl.No.	*Name of Tachinid Fly*	*Jan.*	*Feb.*	*March*	*April*	*May*	*June*	*July*	*Aug.*	*Sept.*	*Oct.*	*Nov.*	*Dec.*
	Sub-family-Tachininae												
	Tribe- Tachinini												
1.	*P. tricolor* sp. n.	–	–	–	–	–	–	–	–	–	+	–	–
	Tribe- Thelairini												
2.	*T. nigripes* F.	–	–	–	–	–	–	–	+Δ	–	+	+	–
	Tribe- Phyllomyini												
3.	*P. panhalaensis* sp.n.	–	–	–	–	–	–	–	–	–	–	–	–
	Sub-family-Exoristinae												
	Tribe-Bloneliini												
4.	*U. uramyoides* Town.	–	–	–	–	–	–	–	–	–	–	–	–
5.	*C. concinnata* (Meigen, 1824).	–	–	–	–	–	–	–	–	+	–	–	–
	Tribe- Exoristini												
6.	*C. dorsalis* sp. n.	–	–	–	–	–	–	–	–	+	–	–	–
7.	*E. larvarum*	–	–	–	–	–	–	–	–	–	–	+	–
	Tribe- Winthemiini												
8.	*W. chandagadensis* sp. n.	–	–	–	–	–	–	–	–	–	–	+	–
	Tribe –Carceliini												
9.	*C. (Carcelia) orientalis* sp.n.	–	–	–	–	–	–	–	–	–	+ Δ	+	–
10.	*C. (Carcelia) sataraensis* sp. n	–	–	–	–	–	–	–	–	–	–	–	–
11.	*C. (Carcelia) abdominalis* sp.n.	–	–	–	–	–	–	–	+	–	–	–	–
12.	*C.* (*Senometopia*) *kolhapurensis* sp. n.	–	–	–	–	–	–	–	–	–	–	–	–

Contd...

Table 2–*Contd...*

Sl.No.	*Name of Tachinid Fly*	*Jan.*	*Feb.*	*March*	*April*	*May*	*June*	*July*	*Aug.*	*Sept.*	*Oct.*	*Nov.*	*Dec.*
	Tribe- Sturmini												
13.	*B. schineri* Mensil	–	–	–	–	–	–	+	–	–	–	–	–
	Tribe- Goniini												
14.	*L. ciliata* (Macquart, 1848)	–	–	–	–	–	–	–	–	–	–	+	–
	Tribe- Eryciini												
15.	*A. striatalis* sp. n	–	–	–	–	–	–	–	–	+	–	+	–
16.	*P.carbonaria* sp.n.	–	–	–	–	–	–	–	–	–	–	–	–
	Sub-family- Dexiinae												
	Tribe- Dexiini												
17.	*P. siberita* F.	–	–	–	–	–	–	–	–	–	–	+	–
	Tribe- Dexia												
18.	*D. vacua* Fall.	–	–	–	–	–	–	–	–	–	–	–	–
19.	*D. indica* sp,n	–	–	–	–	–	–	–	–	–	–	–	–
	Sub-family- Phasiinae												
	Tribe- Cylindromyini												
20.	*C. scutellata* sp.n.	–	–	–	–	–	–	–	–	–	–	–	–

Table 3: Seasonal Abundance of Tachinid Flies from Satara District (2011-2015)

Sl.No.	*Name of Tachinid Fly*	*Jan.*	*Feb.*	*March*	*April*	*May*	*June*	*July*	*Aug.*	*Sept.*	*Oct.*	*Nov.*	*Dec.*
	Sub-family-Tachininae												
	Tribe- Tachinini												
1.	*P. tricolor* sp. n.	–	–	–	–	–	–	–	–	–	–	+	–
	Tribe- Thelairini												
2.	*T. nigripes* F.	–	–	–	–	–	–	–	–	+	+	–	–
	Tribe- Phyllomyini												
3.	*P. panhalaensis* sp.n.	–	–	–	–	–	–	–	–	–	–	–	–
	Sub-family-Exoristinae												
	Tribe-Bloneliini												
4.	*U. uramyoides* Town.	–	–	–	–	–	–	–	–	+	–	–	–
5.	*C. concinnata* (Meigen, 1824).	–	–	–	–	–	–	–	–	–	–	–	–
	Tribe- Exoristini												
6.	*E. larvarum*	–	–	–	–	–	–	–	–	–	–	+	–
7.	*C. dorsalis* sp. n.	–	–	–	–	–	–	–	–	+	+	–	–
	Tribe- Winthemiini												
8.	*W. chandagadensis* sp. n.	–	–	–	–	–	–	–	–	–	–	–	–
	Tribe -Carceliini												
9.	*C. (Carcelia) orientalis* sp.n.	–	–	–	–	–	–	–	–	–	+ Δ	+	–
10.	*C. (Carcelia) sataraensis* sp. n	–	–	–	–	–	–	–	–	–	+	–	–
11.	*C. (Carcelia) abdominalis* sp.n.	–	–	–	–	–	–	–	+	–	–	–	–
12.	*C.* (*Senometopia*) *kolhapurensis* sp. n.	–	–	–	–	–	–	–	–	–	+	–	–

Contd...

Table 3–*Contd...*

Sl.No.	*Name of Tachinid Fly*	*Jan.*	*Feb.*	*March*	*April*	*May*	*June*	*July*	*Aug.*	*Sept.*	*Oct.*	*Nov.*	*Dec.*
	Tribe- Sturmini												
13.	*B. schineri* Mensil	–	–	–	–	–	–	–	–	–	+	+ Δ	–
	Tribe- Eryciini												
14.	*A. striatalis* sp. n	–	–	–	–	–	–	–	+ Δ	+	–	+	–
15.	*P. carbonaria* sp.n	–	–	–	–	–	–	–	–	–	–	–	–
	Tribe- Goniini												
16.	*L. ciliata* (Macquart, 1848)	–	–	–	–	–	–	–	–	–	–	–	–
	Sub-family- Dexiinae												
	Tribe- Dexiini												
17.	*P. siberita* F.	–	–	–	–	–	–	–	–	–	+	–	–
	Tribe- Dexia												
18.	*D. vacua* Fall.	–	–	–	–	–	–	–	–	–	+	–	–
19.	*D. indica* sp,n	–	–	–	–	–	–	–	–	–	–	–	–
	Sub-family- Phasiinae												
	Tribe- Cylindromyini												
20.	*C. scutellata* sp.n.	–	–	–	–	–	–	–	–	–	–	–	–

Talbe 4: Seasonal Abundance of Tachinid Flies from Sindhudurg District (2011-2015)

Sl.No.	*Name of Tachinid Fly*	*Jan.*	*Feb.*	*March*	*April*	*May*	*June*	*July*	*Aug.*	*Sept.*	*Oct.*	*Nov.*	*Dec.*
	Sub-family-Tachininae												
	Tribe- Tachinini												
1.	*P. tricolor* sp. n.	–	–	–	–	–	–	–	–	–	–	–	–
	Tribe- Thelairini												
2.	*T. nigripes* F.	–	–	–	–	–	–	–	+	–	–	–	–
	Tribe- Phyllomyini												
3.	*P. panhalaensis* sp.n.	–	–	–	–	–	–	–	–	–	–	–	–
	Sub-family-Exoristinae												
	Tribe-Bloneliini												
4.	*U. uramyoides* Town.	–	–	–	–	–	+	–	–	–	–	–	–
5.	*C. concinnata* (Meigen, 1824).	–	–	–	–	–	–	–	–	–	–	–	–
	Tribe- Exoristini												
6.	*C. dorsalis* sp. n.	–	–	–	–	–	–	–	–	–	–	–	–
7.	*E. larvarum*	–	–	–	–	–	–	–	–	+	–	+	–
	Tribe- Winthemiini												
8.	*W. chandagadensis* sp. n.	–	–	–	–	–	–	–	–	–	–	–	–
	Tribe -Carceliini												
9.	*C. (Carcelia) orientalis* sp.n.	–	–	–	–	–	–	–	+	–	–	–	–
10.	*C. (Carcelia) sataraensis* sp. n	–	–	–	–	–	–	–	–	–	–	+	–
11.	*C. (Carcelia) abdominalis* sp.n.	–	–	–	–	–	–	–	–	+	–	–	–
12.	*C.* (*Senometopia*) *kolhapurensis* sp. n.	–	–	–	–	–	–	+	–	–	–	–	–

Contd...

Table 4–*Contd...*

Sl.No.	*Name of Tachinid Fly*	*Jan.*	*Feb.*	*March*	*April*	*May*	*June*	*July*	*Aug.*	*Sept.*	*Oct.*	*Nov.*	*Dec.*
	Tribe- Sturmini												
13.	*B. schineri* Mensil	–	–	–	–	–	–	–	–	–	–	–	–
	Tribe- Eryciini												
14.	*A. striatalis* sp. n	–	–	–	–	–	–	–	–	+	–	–	–
15.	*P.carbonaria* sp.n.	–	–	–	–	–	–	–	–	–	–	–	–
	Tribe- Goniini												
16.	*L. ciliata* (Macquart, 1848)	–	–	–	–	–	–	–	–	–	–	+	–
	Sub-family- Dexiinae												
	Tribe- Dexiini												
17.	*P. siberita* F.	–	–	–	–	–	–	–	–	–	–	+Δ	–
	Tribe- Dexia												
18.	*D. vacua* Fall.	–	–	–	–	–	–	–	–	–	–	–	–
19.	*D. indica* sp,n	–	–	–	–	–	–	–	–	–	–	–	–
	Sub-family- Phasiinae												
	Tribe- Cylindromyini												
20.	*C. scutellata* sp.n.	–	–	–	–	–	–	–	–	–	–	–	–

A total of 12 species were reported from Satara district (Table 3). *T. nigripes, C. orientalis, D. vacua* and *B. schineri* showed relatively high population than others. From Satara district Mahabaleshwar, Koyna and Patan are near to Western Ghats which were good spots for tachinid population. Average population of tachinids was reported from Phaltan, Wai, Sajjangad, Dahiwadi and Mhaswad. Population peak was observed maximum during September, October and November. Low population was observed during months of January to May because of summer and less availability of host species. The dominant sub-family from this district was Exoristinae while, the sub-family Phasiinae was absent (Figure 226).

Table 4 shows distributional pattern and seasonal abundance of tachinid flies in Sindhudurg district. Out of 20 species, 08 species were reported from Sindhudurg district. From reported species *P. siberita* showed maximum population. From Sindhudurg district Devgarh is near to Western Ghats which is good spot for tachinid population. Minimum diversity of tachinids was reported from Kankawali, Achra, Malwan, Kudal, Vengurla and Amboli. Population peak was observed during July and November. Low population was observed because of summer and less availability of host species during months of January to May. The dominant sub-family from this district was Exoristinae while, the sub-family Phasiinae was absent (Figure 226).

From the comparative study of tachinid species from selected four districts it is concluded that, maximum diversity of tachinids was reported from Kolhapur district and minimum diversity of tachinids was reported from Sangli and Sindhudurg district. Months of July to November were good season for collection of tachinid flies. Two species of tachinids *i.e. T. nigripes, C. orientalis* and *P. siberita* reported from selected districts. In general the forest ecosystems were dominated with all four subfamilies than that of the agroecosystems (Figures 227 and 228).

Discussion

Lahiri (2003) reported 15 tachinid species from the Sikkim state of India. Out of which two were newly described *viz., Esteria intermedia* and *Eurithia indica*. Another nine species were recorded for the first time from Sikkim state. Lahiri (2006) reported 15 tachinid species from the Nagaland state of India. Out of which he added four new species from the region *viz., Dexia prakritae, Dexia quadristriata, Paradrino pilifacies* and *Phorinia nigra*. Another nine species were recorded for the first time from Nagaland state by Lahiri (2006).

Dawah (2011) recorded ten species of tachinid flies from south western Saudi Arabia. Eight of which were reported for the first time indicating the occurrence of 10 tachinids from Palaearctic and Afrotropical regions.

In the present work, 20 tachinids have been reported from Kolhapur, Sangli, Satara and Sindhudurg districts of Western Maharashtra, India as additional species reported than Sathe (2012). The species reported from this region were *W. chandgadensis sp.n., P. tricolor sp.n., D. indica sp.n., C. orientalis sp.n, C. sataraensis sp.n., S. kolhapurensis sp.n., C. scutellata sp.n., A. striatalis sp.n., C. ruralis sp.n., P. panhalaensis sp.n., P.carbonaria* sp.n. and *C. abdominalis* sp.n.

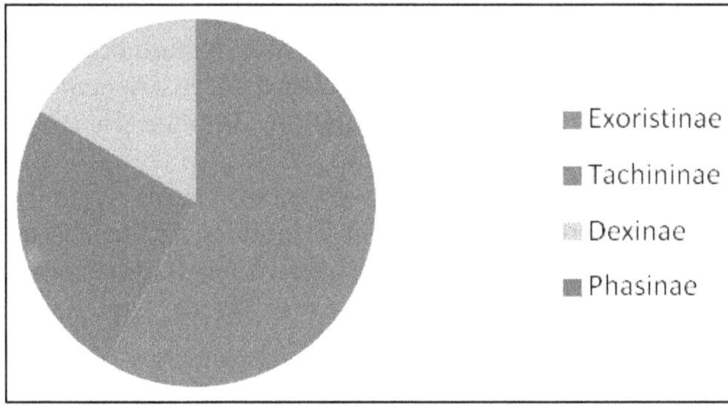

Figure 227: Occurrence of Subfamilies of Tachinids from Agroecosystem of Western Maharashtra.

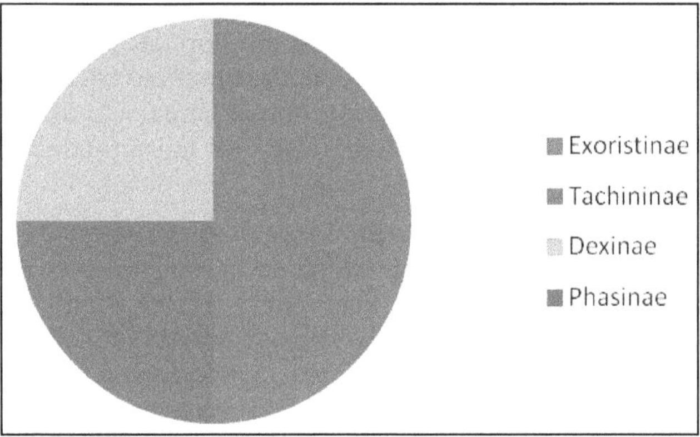

Figure 228: Occurrence of Subfamilies of Tachinids from Forest Ecosystem of Western Maharashtra.

Inclan and Stireman, III (2011) studied seasonal abundance and distributional record of tachinid flies for 2 yrs from southwestern Ohio. During survey 883, tachinid specimens were collected, consisting of 117 species belonging to 69 genera. The majority of the specimens were collected in rainy season, followed by summer and spring, with 67, 22, and 11 per cent, respectively. Distributional record of 217 species showed that little diversity of Tachinidae occurred from selected areas *i.e.* 16 and 19 per cent from northeastern United States and Ohio respectively. In North America very less diversity was seen. Results help to understand the temporal and spatial diversity of ecologically and agronomically important parasitoids.

O'Hara (2011) studied world genera of the tachinidae and their regional occurrence. In their studies a total of 9899 species were collected from Neotropical, Nearctic, Palaearctic, Afrotropical, Oriental and Australasian regions of the world. The availability of specimen from selected region was 78 per cent, 20 per cent, 35 per cent, 43 per cent, 18 per cent and 55 per cent respectively. He concluded that, there are nearly 10,000 described species of tachinidae in the world but, their number is more because the Neotropical, Afrotropical, Oriental and Australasian regions contain large numbers of undescribed species.

Sathe (2012) studied the biodiversity of tachinid flies from western Maharashtra. He reported 20 species attacking various lepidopterous pests including *Helicoverpa armigera* (Hubn.), *Spodopetra litura* Fab., *Achea janata* (Linn.) and *Tarache notabilis* Walk. Shendge and Sathe (2014) studied the biology of *Eucelatoria bryani* Sabr. (Diptera: Tachinidae), a larval parasitoid of *H. armigera* (Lepidoptera: Nocuidae).

Sathe et. al. (2014) studied tachinid flies with respect to floral host plants and their sex ratio. A total of 82 species were collected from 13 floral plants from Kolhapur and Satara districts. Out of 13 floral host plants, *Tecoma castanifolia* and *Eupatorium odoratum* plants showed maximum association of tachinids with them. Sathe *et al.* (2014) also concluded that in Kolhapur district availability of tachinids was more than Satara district.

In the present work, Seasonal abundance, distributional record, host plants and sex ratio of 17 tachinid species was studied from Kolhapur, Sangli, Satara and Sindhudurg districts of Western Maharashtra, India. Maximum diversity of tachinid flies was reported from Kolhapur district and minimum diversity of tachinids was reported from Sangli and Sindhudurg district. Months of July to November were good duration for collection of tachinid flies from every district. Two species of tachinids *i.e. T. nigripes* and *P. siberita* reported from all four selected districts. Availability of tachinid species will be useful for biological control of insect pest from agroecosystems as well as forest ecosystems. The dominant sub-family found in western Maharashtra was Exoristinae. However, the sub-family Phasiinae was not recorded from Western Maharashtra except Kolhapur district. The present work will be helpful for exploiting tachinid fauna in biological control of insect pests.

6
Summary and Conclusion

Maharashtra is richest state for floral and faunal diversity from India. Many endemic, rare, indeterminate and endangered species of animals and plants are found in Western Ghats of Maharashtra. However, very little information is available of insect's biodiversity from western Maharashtra. As regards to tachinids very scanty literature is available from Maharashtra. Hence, the present work was carried out from Western Maharashtra especially from Kolhapur, Sangli, Satara and Sindhudurg districts.

Tachinid flies belongs to order Diptera of Class-Insecta containing 10, 000 described species all over the world. Tachinid flies can be distinguished by small size, presence of setae and sucking type of mouth parts. Tachinids are important parasitoids of many insects which helps to control notorious insect pests by acting as biocontrol agents.

The book has been divided into five chapters. First chapter is devoted for a general introduction that explains importance of tachinid flies and national and international status of topic.

Second chapter is devoted to review of literature on taxonomy and seasonal abundance of tachinids.

The chapter third deals with materials and methods adopted for completion of present work. The tachinid flies have been collected with the help of insect collecting net. The collected specimen have been pinned and preserved in storage box and taxonomical studies have been made on them. For taxonomical studies insect parts like head, thorax, abdomen and their appendages have been taken into account. The collection was done by visiting study spots at 15 days interval by one man one hour search method. Seasonal abundance and distributional record of tachinid flies was made by spot observations at different study spots. The host plants and tachinids were identified by consulting appropriate literature.

Fourth chapter embodies the taxonomy of tachinids. The taxonomical studies include description of 12 new species of tachinids namely, *W. chandgadensis sp.n., P. tricolor sp.n., D. indica sp.n., C. orientalis sp.n, C. sataraensis sp.n., S. kolhapurensis sp.n., C. scutellata sp.n., A. striatalis sp.n., C. dorsalis sp.n., P. panhalaensis sp.n., P.carbonaria* sp.n. and *C. abdominalis* sp.n. and redescription of 8 species namely, *T. nigripes, U. uramyiodes, B. schineri, P. siberita, E. larvarum, C. compsilura, L.ciliata* and *D. vacua.* Seasonal abundance and distributional record of tachinid flies have also been included under separate chapter. A total of 20 species of tachinids have been reported from western Maharashtra. The dominant families of the region refer to Exoristinae followed by Tachiniae and Dexiinae. However, the sub-family Phasininae was only found in Kolhapur district. The chapter sixth deals with summary and conclusion of the work. Finally bibliography of the topic is appended at last.

Conclusion

Tachinids are good biocontrol agents of several types of insect pests including Lepidoptera, Orthoptera, Coleopteran, etc. However, this group of insects is largely neglected from India. Therefore, in the present study 20 species of tachinids were described belonging to four sub-families, 12 tribes, 13 genera and 1 sugenera from Kolhapur, Sangli, Satara and Sindhudurg districts of Western Maharashtra. 10 new species namely *W. chandgadensis sp.n., P. tricolor sp.n., D. indica sp.n., C. orientalis sp.n, C. sataraensis sp.n., S. kolhapurensis sp.n., C. scutellata sp.n., A. striatalis sp.n., C. dorsalis sp.n., P. panhalaensis sp.n. P.carbonaria* sp.n. and *C. abdominalis* sp.n. have been described for the first time and 5 species *T. nigripes, U. uramyiodes, B. schineri, P. siberita* and *D. vacua* have been redescribed. Out of 20 species, 4 species were rare and 13 were common.

Seasonal abundance and distribution of tachinids in the region indicates that, out of four sub-families Exoristinae was more dominant over other subfamilies. The members of sub-family Exoristinae were abundantly scattered in agro and forest ecosystems. From sub-family tachininae, *U. uramyoides* reported for the first time from India.

Insect pests cause severe destruction to agro and forest crop plants from Maharashtra. Tachinidae are good biocontrol agents found in agro and forest ecosystems. The tachinids are biopesticides scattered in the region of western Maharashtra. They should be identified, conserved, protected and used for pest management as ecofriendly tool.This work will be stimulatory to workers in the field of eco-tfriendly (biocontrol) pest management.

References

Adamo, S. A., D. Robert, J. Perez and Hoy. R. R. 1995b. The response of an insect parasitoid, *Ormia ochracea* (Tachinidae), to the uncertainty of larval success during infestation. *Behav. Ecol. Sociobiol.*, 36: 111- 118.

Adamo, S.A., D. Robert, J. Perez and Hoy, R. R. 1995a. Effects of a tachinid parasitoid, Ormiaochracea, on the behavior and reproduction of its male and female ûeld crickethosts (*Gryllus* spp.). *Journal of Insect Physiology*, 41: 269-277.

Aldrich, J. M. 1925. The two new species of the tachinid genus *Lixophaga* with notes and key (Diptera) *Proc. Ent.Soc. Wash.*, 27: 132-136.

Aldrich, J. M. 1931. Notes on the genus Cheatonodexodes, with one new species. *Ann. Mag. Nat.Hist.*, 10 (8): 205-207.

Arnaud, P. H. 1963. *Perumyia embiaphaga, a* new genus and species of Neotropical Tachnidae parasitic on Embioptera. *Am. Mus. Novit.*, 2143: 1- 9.

Baranov, N. 1932. Larvaevoridae (Ins. Dipt.) von Sumatra, I. *Miscnea zool. Sumatr.*66:1-3.

Baranov, N. 1932. Neue orientalische Tachinidae. *Encycl.ent.*(B) II, 6:83-93.

Barraclough, D. A. 2005. *Rhinophoroides minutus*, a new genus and species of rare nocturnal Dufouriini (Diptera: Tachinidae: Dexiinae) from South Africa. *African Entomology.*, 13: 380–384.

Beason, C.F. 1941. Ecology and control of forest insects of India and neighbouring countries, Govt. of India (1961 Reprint), pp. 767.

Belshaw, R. 1994. Life history characteristics of Tachinidae (Diptera) and their effect on polyphagy, pp. 145-162. In Hawkins B. and Sheehan W. [eds.], Parasitoid community ecology. *Oxford University Press, New York, USA*, 516 p.

Beneway, D. F. 1963. Revision of the flies from the genus *Lespesia* in North America (Diptera: Tachinidae). *Kans. Univ. Sci. Bull.*, 44: 627-686.

Bentley, H. and Raper, C. 2010. *Linnaemya picta* (Meigen, 1824) (Diptera, Tachinidae) new to Britain. *Dipterists Digest.*, 17: 77–78.

Bergstrom, C. 2005. A new species of the genus *Phebellia* Robineau-Desvoidy (Diptera: Tachinidae) from Finland. Stuttgarter Beiträge zur Naturkunde Serie A (Biologie). 679: 1-9.

Bergstrom, C. 2007. A new species of *Nilea* Robineau-Desvoidy (Diptera, Tachinidae) with notes on the genus and a key to the North European species. *Entomologisk Tidskrift* 128: 151–161.

Bergstrom, C. 2007. *Loewia erectan.* sp. (Diptera: Tachinidae)- a new parasitic fly from Fennoscandia and Poland. Stuttgarter Beiträge zur Naturkunde Serie A (Biologie) 708: 1–16.

Bhoje, P.M. and T.V.Sathe, 2003. Faunastic studies on butterflies from Radhanagari Wildlife Sanctuary, India. *Indian J. Environ. and Ecoplan.*, 7(3), 655-658.

Brooks, A. R. 1943. Reviewed the Canadian species of *Ernestia Sens* Lat. (Diptera: Tachinidae). *Can. Ent.*, 75: 66-78.

Bushbeck, E. K., and Strausfeld, N. J. 1997. The relevance of neural architecture to visual performance: Phylogenetic conservation and variation in dipteran visual systems. *J. Comp. Neurol.*, 383: 282–304.

Bystrowski, C. 2001. A new species of the genus *Campylocheta* Rondani, 1859 (Diptera: Tachinidae) from Poland. Annales Zoologici (Warszawa), 51(3): 279–281.

Bystrowski, C. 2006. *Ramonda latifrons* (Zetterstedt, 1844) (Diptera: Tachinidae) – a tachinid fly new to Polish fauna.[In Polish.] *Wiadomosci Entomologiczne*, 25: 179–182.

Byun, H. W. and Han, H. Y. 2009. A taxonomic revision of the genus *Metadrinomyia* Shima (Diptera: Tachinidae) with descriptions of two new species. *Zootaxa.*, 2311: 55–65.

Cade, W. H. 1975. Acoustically orienting parasitoids: y phonotaxis to cricket song. *Science.*, 190: 1312-1313.

Campbell, H. R. and Strausfeld, N. J. 2001. Learned discrimination of pattern orientation in walking ies. *J. Exp. Biol.*, 204: 1–14.

Cantrell, B. K. and Burwell, C. J. 2010. The tribe Dufouriini (Diptera: Tachinidae: Dexiinae) recorded from Australia with the description of two new species. *Memoirs of the Queensland Museum.*, 55: 119–133.

Cerretti P., De Biase, A. and Freidberg, A. 2009. Systematic study of the genus Rossimyiops Mesnil (Diptera, Tachinidae). *Zootaxa*, 1984: 31–56.

Cerretti, P. 2009. A new Afrotropical genus of Voriini, with remarks on related genera (Diptera: Tachinidae: Dexiinae). *Insect Systematics and Evolution.*, 40: 105–120.

Cerretti, P. and Barraclough, D.A. 2007. *Anomalostomyia namibica,* a new genus and species of Afrotropical Tachinidae (Diptera). *Italian Journal of Zoology* 74: 101–106.

Cerretti, P. and Tschorsnig, H. P. 2012. Three new species of *Estheria* Robineau-Desvoidy (Diptera: Tachinidae) from the Mediterranean, with a key to the European and Mediterranean species of the genus. *Stuttgarter Beiträgezur Naturkunde A (Biologie),* N. Ser. 5: 271–286.

Cerretti, P. and Tschorsnig, H.-P. 2007. Two new species of *Siphona* Meigen (Diptera: Tachinidae) from Sardinia and Morocco. *Stuttgarter Beiträge zur Naturkunde. Serie A* (Biologie), 704: 1–7.

Cerretti, P. and Tschorsnig, H.-P. 2008. A new species of *Plesina* Meigen (Diptera: Tachinidae) from the Medi-terranean. *Stuttgarter Beiträge zur Naturkunde A, New Ser*. 1: 445–450.

Cerretti, P. and Wyatt, N. 2006. A new species of *Eomedina* Mesnil (Diptera: Tachinidae) from Namibia. *Zootaxa,* 1147:61-68.

Cerretti, P., Tschorsnig H.-P., and Lopresti, M. 2010. Moschweb, interactive key to the Palaearctic genera of Diptera: Tachinidae,http://www.tachinidae.eu (accessed December 2010).

Cerretti, P., Wood, D.M. and O'Hara, J.E. 2012. *Neoethilla,* a new genus for the first record of the Ethillini from the New World (Diptera, Tachinidae, Exoristinae). *ZooKeys,* 242: 25–41.

Cha, D.-j. and Han, H.-y. 2009. Two species of the genus *Phasia* Latreille (Insecta: Diptera: Tachinidae) new to Korea. *Korean Journal of Systematic Zoology,* 25: 309–316.

Champion, H.G. and S.K. Seth 1968. A revised survey of the forest type of India. Manager of Publication, Delhi, pp. 464.

Chao, C.M. 1962. Fauna Larvaevoriden Chinas. V. gattung *Exorista* Meigen [In Chinese with German descriptions of new species]. *Acta Ent. Sin.,* 13: 362-375.

Chao, C.-m. 1976. New species of the genus *Thecocarcelia* T.T. from China (Diptera: Tachinidae) [In Chinese with Russian key]. *Acta Entomologica Sinica,* 19: 335– 338.

Chao, C.-m. 1979. New species of the subtribe Peleteriina from China (Diptera: Tachinidae) [In Chinese with English summary]. *Acta Zootaxonomica Sinica,* 4: 156–161.

Chao, C.-m. and Chen, X.-l. 2007. A taxonomic study on the genus Phebellia R-D (Diptera: Tachinidae) from China [In Chinese]. *Acta Entomologica Sinica,* 50: 933–940.

Chao, C.m. and Liang, E.y. 2002. Review of the Chinese *Carcelia* Robineau-Desvoidy (Diptera: Tachinidae) [In Chinese]. *Acta Zootaxonomica Sinica,* 27: 807-848.

Chao, C.m. and Liang, E.y. 2003. A study of the Chinese genus *Smidtia* Robineau-Desvoidy (Diptera, Tachinidae) [In Chinese]. *Acta Zootaxonomica Sinica,* 28: 152-158.

Chao, C.-m. and Shi, Y.-s. 1980. Descriptions of the genus *Blepharipa* Rondani of China (Diptera: Tachinidae). Pp.1–6. In: Liaoning Institute of Sericulture, ed., collected works of *Blepharipa* Rondani [In Chinese]. Science Press, Beijing. 139 pp.

Chao, C.-m. and Shi, Y.-s. 1980. Notes on new species of *Hyalurgus* Brauer and Bergenstamm (Diptera: Tachinidae) [In Chinese with English summary]. *Acta Entomologica Sinica*, 23: 314–321.

Chao, C.-m. and Zhou S.-x. 1987. New species of tachinid flies from Hengduan Mountains of China (Diptera: Tachinidae) [In Chinese with English summary]. *Sino-zoologia*, 5: 207–215.

Chao, C.-m., Sun, X.-k. and Zhou, S.-x. 1990. Studies on the tribe Parerigonini from China (Diptera: Phasiinae). [In Chinese with English summary.] Acta Zootaxonomica Sinica 15: 230–241.

Chatterjee, P.N. and M.P. Misra, 1975. Natural insect enemy and plant host complex of forest insect pests of Indian region. *Indian For. Bull. No. 265* (M.S.) Entomology, Controller of Publications, Govt. of India, Delhi pp. 1-223.

Chi, Y., Hao, J. and Zhang, C.-t. 2010. Taxonomic study of the subgenus *Servillia* Robineau- Desvoidy (Diptera, Tachinidae) from North China. Journal of Shenyang Normal University (Natural Science) 28: 86–89. [In Chinese with English summary.].

Collins, G.A., Gibbs, D.J. and Raper, C.M. 2002. *Paracraspedothrix montivaga* Villeneuve and *Carcelia bombylans* Robineau-Desvoidy (Diptera, Tachinidae) new to Britain. *Dipterists Digest* (2nd series) 9: 61-64.

Crosskey R.W. 1976: A taxonomic conspectus of the Tachinidae (Diptera) of the Oriental region. — *Bulletin of the British Museum (Natural History) Entomology*, Supplement 26: 1-357.

Crosskey R.W.1973. A conpectus of the Tachinidae (Diptera) of Australia, including keys to to the supraspecific taxa and taxonomic and host catalogues. *Bull.Br. Mus. Nat. Hist.*(Ent.) Suppl.21, 221pp.

Crosskey, R. W. 1967. Two new genera and species of eryciine tachinidae (diptera) from Australia, Commonwealth Institute of Entomology, London, *J. Aust. ent. Soc.*, 6: 27-35.

Crosskey, R. W. 1976. A taxonomic conspectus of the Tachinidae (Diptera) of the Oriental Region. *Bulletin of the British Museum (Natural History) Entomology Supplement*. 26: 357 pp.

Crosskey, R. W. 1977. A review of the Rhinophoridae (Diptera) and a revision of the Afrotropical species. *British Museum (Natural History), London, Entomology*, 36 (1) 66pp.

Das, B.C. 1993. A new species of *Turanogonia* Rogdendorf, (Diptera: Tachinidae) from Darjeeling distt. West Bengal. *Rec. Zool. Surv. India*, 91: 49-51.

Dawah, H. A. 2011. Some Tachinidae (Diptera: Calyptrata) from South- Western Saudi Arabia. *Journal of Jazan University Applied Sciences Branch.*,1(1): 28-38.

Dethier, V. G. 1963.The Physiology of Insect Senses, *J. Wiley, New York.*

Diego J. Inclan and John O. Stireman, III. 2011. Tachinid (Diptera: Tachinidae) Parasitoid Diversity and Temporal Abundance at a Single Site in the Northeastern United States. *Annals of the Entomological Society of America* 104(2):287-296.

Disney, H. 2000. Hands-on taxonomy. *Nature.,* 405: 307.pp 619.

Draber Monko, A. and Nowakowski, E. 2010. The tachinid *Euthera fascipennis* (Loew, 1854) (Diptera, Tachinidae) new to the fauna of the Tunisia, with a description of the female and the puparium. *Fragmenta Faunistica,* 52[2009]: 197–205.

Ebrahim Gilasian, Ali Asghar Talebi, Joachim Ziegler, Shahab Manzari and Mehrdad Parchami Araghi 2013. A review of the genus *Cylindromyia* Meigen (Diptera: Tachinidae) in Iran, with the description of two new species and the newly discovered male of *C. persica* Tschorsnig, *Studia dipterologica,* 20(2): 299–324.

Falk, S.J., Perry, I. and Howe, M.A. 2009. Leucostoma anthracinum (Meigen, 1824) (Diptera, Tachinidae) new to Britain. *Dipterists Digest* (2nd Series) 16: 87–88.

Feener, D. H., Jr., and Brown, B. V. 1997. Diptera as parasitoids. *Annu. Rev. Entomol.,* 42: 73-97.

Ferrar, P. 1987. "A guide to the breeding habits and immature stages of Diptera Cyclorrhapha". Entomonograph: 8: 367-384. *E.J. Brill/Scandinavian Science Press,* Leiden, Copenhagen.

FST, 1996. Testing and developing criteria and indicators for sustainable management of tropical plantation forests. *Project ID FST/1996/187.*

Gammelmo and Sagvolden. 2007. The tachinid fly, *Phasia hemiptera* (Fabricius, 1794) (Diptera, Tachinidae) in Norway. *Norw. J. Entomol.,* 54: 51-54.

Gaonkar, H. 1996. The butterflies of Western Ghats India and Srilanka. A biodiversity assessment on the ecological history of Western Ghats Abst., pp. 99.

Georgiev, G. 2000. New and rare Tachinidae (Diptera) parasitoids of insect pests on the poplars (*Populus* spp.) in Bulgaria. *Nauka za Gorata* 37: 49-56.

Ghahari *et al.,* 2008. A contribution to the dipteran parasitoids and predators in Iranian cotton fields and surrounding grasslands. *Mun. Ent. Zool.* 3(2).

Gheibi, M. and Ostovan, H. 2009. Preliminary investigation on the fauna of tachinid flies (Diptera: Tachinidae) in Fars Province, Iran. *Plant Protection Journal.,* 1: 140–166.

Gheibi, M., Ostovan, H. and Kamali, K. 2009. A contribution to knowledge of the tachinid fly fauna of Fars province, Iran (Diptera: Tachinidae: Dexiinae and Phasiinae). *Zoology in the Middle East.,* 46: 69–74.

Gheibi, M., Ostovan, H., Kamali, K. and Gilasian, E. 2009. The first report of six tachinid flies from Iran (Diptera: Tachinidae). *Journal of Entomological Society of Iran* 29: 53– 55. [In Persian with English abstract.].

Gilasian, E., Talebi, A. A., Ziegler, J. and Manzari, S. 2013. A taxonomic study of the genus Phasia (Dip.: Tachinidae) in Iran, with two new records. *Journal of Entomological Society of Iran.*, 33: 13.

Gilasian, E., Talebi, A. A., Ziegler, J. and Manzari, S. 2013. A review of the genus *Phania* Meigen, 1824 (Diptera: Tachinidae: Phasiinae) in Iran with the description of a new species. *Zoology and Ecology.*, 23: 13–19.

Global Forest Resources Assessment 2010, FAO Forestry Paper 163, Food and Agriculture Organization of the United Nations (2011), pp. 12-13.

Griener, S. 1988. "Applied biological control with tachinid flies (Diptera: Tachinidae): A review". Anzeiger für Schadlingskunde, Pflan-zenschutz. *Umweitschutz*, 61: 49-56.

Groombridge, B. 1992. Global biodiversity: status of the Earths living resources, London.

Guimarâes, J. H. 1977. Host-parasite and parasite-host catalogue of South American Tachinidae (Diptera). *Arq. Zool.*, 28 (3): 1-131.

Hao, J., Zhang, C.-t. and Chi, Y. 2008. Taxonomic study of subgenus *Tachina* Meigen (Diptera, Tachinidae) from North China [In Chinese with English summary]. *Journal of Shenyang Normal University (Natural Science)*, 26: 476– 479.

Herting, B. 1986. Description of a new species of Tachinidae from Israel and remarks. *Israel Journal of Entomology*, 3: 307–313.

Hickman, J. and Wratten, S. 1996. Use of *Phelia tanacetifoliastrips* to enhance biological control of aphids by hoverfly larvae in cereal fields. *Journal of Economic Entomology*, 89: 832–840

Hiroshi, S. 1987. A Revision of the Genus *Isosturmia* Townsend (Diptera, Tachinidae), *Bull. Kitakyushu Mus. Nat. Hist.*, 6: 213-237.

Hiroshi, S. 1996. A systematic study of the genus *Cavillatrix* Richter (Diptera:Tachinidae), *Bulletin of the Graduate School of Social and Cultural Studies*, Kyushu University, 2: 133-148.

Irwin, M. E., Schlinger, E. I. and Thompson, F. C. 2003. Diptera, true flies pp. 692-702. In: Goodman, S.M. and Benstead, J.P. The Natural History of Madagascar, University of Chicago Press, Chicago and London, 1728 pp.

Isaacs, R., Tuell, J., Fielder, A., Gardiner, M. and Landis, D. 2008. Maximizing arthropod mediated ecosystem services in agricultural landscapes, the role of native plants. *Frontiers in Ecology and Environment.*, 7: 196–203.

Jadav Divya K. and Sathe T. V. 2014. Altitudinal diversity of forensic Blow flies (Diptera: Calliphoridae) of Western Ghats (Maharashtra). *J. Forensic Res.*, 5, 251.

Jadav Divya K. and Sathe T. V. 2015. Diversity of forensic Blow flies (Diptera: Calliphoridae) from Western Ghats, Maharashtra, India. *Indian journal of applied research*, 5(9), 55-60.

Jadav Divya K. and Sathe T. V. 2015. Diversity, occurrence and development of forensic insects on Dog (*Canis domesticus* L.) carcass from Kolhapur, India. *International Journal of Pharma & Biological sciences*, 6(4)(B), 498-506.

Jadav, D. and T.V. Sathe 2014. Altitudinal diversity of forensic blowflies (Diptera: Calliphoridae) of Western Ghats (Maharashtra). *J. Forensic Res.*, 5(5): 100-251.

Jadhav A. D., Desai A. S. and T. V. Sathe. 2014. Distribution and economic status of Uzi Fly *Exorista Bombycis* Louis, a parasitoid of mulberry silk worm *Bombyx Mori* L. in Maharashtra. *Global journal for research analysis*, 3 (7), 3- 5.

James, E. O' Hara. 2008. World Genera of the Tachinidae (Diptera) and their Regional Occurrence Version. 4.0.

Jones, R.A. 2006. Another record of *Catharosia pygmaea* (Fallen, 1815) (Diptera, Tachinidae) in Kent. *Dipterists Digest* (2nd series), 13: 100.

Jones, R.A. 2009. *Gymnosoma rotundatum* (Linnaeus, 1758) (Diptera, Tachinidae) established in urban London. *Dipterists Digest* (2nd Series), 16: 81–82.

Joseph, T.M. 2004. Biodiversity conservation in the Western Ghats. *Biodiversity and Environment*, 17, 221-227.

Judd, W. S., Campbell, C. S., Kellog, E. A., Stevens, P. F., Donoghue, M. J. 2007. Taxonomy. In Plant Systematic – A Phylogenetic Approach, Sinauer Associates, Sunderland. 3rd ed.

Kadam and Sathe. 2014. Biosystematics of Lepidoptera from Western Maharshtra including Ghats. Phd Thesis. Shivaji University, Kolhapur, India.

Kara and Aksu, 2008. Contributions to the Turkish Tachinidae (Diptera) Fauna. *Turk J Zool.* 32: 227-228.

Kara, K. 2001. Additions to the fauna of Turkish Tachinidae (Insecta, Diptera). *Zoology in the Middle East.*, 23: 85-88.

Karagoz, M., Aksu, S., Gozüacik, C. and Kara, K. 2011. Microphthalma europaeaEgger (Diptera: Tachinidae), a new record for Turkey. *Turkish Journal of Zoology*, 35: 887–889.

Kavane, R.P. and T.V. Sathe 2011. Wild Silk technology. *Daya publishing house, New Delhi*, pp. 1-234, ISBN- 948-81-7035-712-4.

Kavane, V.V., 2011. Biosystematics of Wild silk moths from western Maharashtra. Phd Thesis. Shivaji University, Kolhapur, India.

Kugler, J. 1972. Tachinidae of Israel V. *Mesnilomyia* and *Palmonia*, two new genera of Tachinidae (Diptera). *Israel Journal of Zoology*, 21: 103–112.

Lahiri, A. R. 2006. Diptera: Tachinidae, In: Fauna of Nagaland, State Fauna series, *Zoological Survey of India*, Kolkata, 12: 199-211.

Lahiri, A.R. 2003. Diptera: Tachinidae. In: Fauna of Sikkim, State Fauna series, Zoological Survey of India, Kolkata, (Part- 3): 387-399.

Le Peletier, A.L. M. Serville, J.G.A. 1828. In Latreille, P.A. *et al., Encyclopedie methodique. Historie naturelle. Entomologie, ou Histoire naturelle des Crustaces, des Arachnides et des Insectes* 10:345-832.

Lee, H.-s. and Han, H.-y. 2008. A taxonomic study of the genus *Frontina* Meigen (Diptera: Tachinidae) in Korea. *Journal of Asia-Pacific Entomology*, 11: 137–143.

Legaspi, J. C., Legaspi, B. C. Jr., King, E. G. and Saldana, R. R. 1997. Mexican rice borer, *Eoreuma loftini* (Lepidoptera:Pyralidae) in the Lower Rio Grande Valley of Texas: its history and control. *Subtropical Plant Science*, 49: 53-64.

Liang, E.-y. and Chao, C.-m. 1992. On the genus *Neophryxe* Townsend from China (Diptera: Tachinidae) [In Chinese with English summary]. *Acta Zootaxonomica Sinica*, 17:224–226.

Liang, E.-y. and Chao, C.-m. 1995. Two new species of the genus *Istochaeta* Rondani from China (Diptera: Tachin-idae) [In Chinese with English summary]. *Acta Zoo taxonomica Sinica*, 20: 487–491.

Lim, J.-s. and Han, H.-y. 2008. Redescriptions of two closely resembling *Linnaemya* species (Insecta: Diptera: Tachinidae) new to Korea. Korean Journal of Systematic Zoology 24: little known species. Israel Journal of Entomology 19(1985): 85–88.

Liu, J. and Zhang, C. T. 2007. A new species of genus Meigenia from China (Diptera,Tachinidae). *Acta Zoo taxonomica Sinica.*, 32 (1): 121- 123.

Liu, J.-y, Zhang, C.-t, Ge, Z.-p and Wang, Y. 2006. Taxonomic study on the tribe Blondeliini from China (Diptera-Tachinidae). *Journal of Shenyang Normal University (Natural Science)*, 24: 334–339.

Liu, Y.-z., Chao, C.-m. and Li, L.-f. 1999. Five new species of Tachinidae from Shanxi province, China (Diptera) [In Chinese with English summary]. *Acta Zootaxonomica Sinica*, 24: 347–354.

Liu, Y.-z., Li, L.-f. and Chao, C.-m. 1986. Descriptions of five new species of Exoristinae (Diptera: Tachinidae) [In Chinese with English summary]. *Entomotaxonomia*, 7(1985): 165–173.

Macquart, J.1834. Insectes Dipteres du nord de la France. Athericeres: Creophiles, Ostrides, Myopaires, Conopsaires, Scenopilinies, Cephalopsides, *Mem. Soc. Sci.Agri. Lille*: 137-368.

Mao, Z.-h. and Chao, C.-m. 1990. A new species of the genus *Parasetigena* from China (Diptera: Tachinidae) [In Chinese with English summary]. *Sinozoologia*, 7: 301- 302.

Martin, W.R. Jr., Nordlund, D. A., and Nettles, W.C. 1990. Response of the parasitoid Eucelatoria bryani to selected plant material in an olfactometer. *J. Chem. Ecol.*, 16: 499–508.

Meigen, J.W. 1803. *Versuch einer neun Gattungs Eitheilung der europaischen zweiflugligen Insekten.* Magazin Insektenk. (Illiger), 2: 259-281.

Meigen, J.W. 1806. *Systematische Beschreibung der bennten europaischen zweiflugeligen Insekten.* 5, xii+412pp. Hamm.

Meigen, J.W. 1953. Larvaevorinae (Tachiininae). *In* Linder, *Fliegen palaearkt.* Reg. 64g: 257-304.

Meigen, J.W. 1963. Nouveaux Tachiniaires de la region palearctiquae principalement de I, URSS et du Japon. *Bull. Inst. R. Sci. nat. Bel.*39 (24), 26pp.

Meigen, J.W. 1966. Larvaevorinae (Tachiininae). *In* Linder, *Fliegen palaearkt.* Reg. 64g: 881-928.

Meigen, J.W. 1973. Larvaevorinae (Tachiininae). *In* Linder, *Fliegen palaearkt.* Reg. 64g: 1113-1232.

Monteith, L.G. 1955. Host preferences of *Drino bohemica* Mesn. (Diptera: Tachinidae) with particular reference to olfactory responses. *Can. Entomol.*, 87: 509–530.

Monteith, L.G. 1956. Inuence of host movement on selection of hosts by *Drino bohemica* Mesn. (Diptera: Tachinidae) as determined in an olfactometer. *Can. Entomol.*, 88: 583-586.

Morrison, F. O. 1940. A revision on American species *Gonia meigen* (Diptera: Tachinidae) *Can. J. Res.*, 18: 336-362.

Narendran, T. C. 2000. The Importance of Systematics. *Resonance.*, 5: 60-68.

Nihei, S. S. and Toma, R. 2010. Taxonomic notes on *Borgmeiermyia* Townsend (Diptera,Tachinidae) with the first host record for the genus. *ZooKeys.*, 42: 101- 110.

Nihei, S.S. 2006. Revision and systematic placement of Prospalaea Aldrich (Diptera, Tachinidae). *Papéis Avulsos de Zoologia* 46: 197–201.

Nihei, S.S. and Pansonato, M.P. 2006. Revision of Prophor-ostomaTownsend, 1927 (Diptera, Tachinidae, Dexiinae), with the description of a new species. Papéis Avulsos deZoologia, 46: 239–244.

Nihei, S.S. and Pavarini, R. 2011. Taxonomic redescription and biological notes on *Diaugia angusta* (Diptera, Tachinidae): parasitoid of the palm boring weevils *Metamasius ensirostris* and *M. hemipterus* (Coleoptera, Dryophthoridae). *ZooKeys.*, 84: 23-38.

Nilam Shendge and Sathe T. V. 2014. B. Biology of *Eucelatoria bryani* Sabr. (Diptera: Tachinidae): A Larval parasitoid of *Helicoverpa armigera* Hubn. (Lepidoptera; Nocuidae). *Recent trends in biological pest control.* Astral Publication, New Delhi, 132- 134.

Norman Myers, Russell A. Mittermeier, Cristina G. Mittermeier, Gustavo A. B. da Fonseca and Jennifer Kents 2000. Biodiversity hotspots for conservation priorities. *Nature*,403:853-858.

Novotná, H., Vaòhara, J., Tóthová, A., Muráriková, N., Bejdák, P. and Rozkošný, R. 2009. Identification and taxonomy of the West Palaearctic species of Tachina Meigen (Tachinidae, Diptera) based on male terminalia and molecular analyses. *Entomologica Fennica*, 20: 139–169.

Nunez, E. and Couri, M.S. 2012. UruleskiaTownsend (Diptera, Tachinidae): redescription of the type-species, description of new species and key to identification. *Papéis Avulsos de Zoologia*, 52: 93–102.

O'Hara, J. 2010."Tachinidae Resources.http://www. nadsdiptera.Org/Tach/ home. htm (accessed October 2010).

O'Hara, J. E. 1983. A new species of siphona (Diptera: Tachinidae) from Australia. *International Journal of Entomology.,*.25(1) : 79-83.

O'Hara, J. E. 1984. *Baeomyia* n.g. (Diptera: Tachinidae): descriptions and notes about phylogenetic and zoogeographic relationships. *Can. J. Zool.*, 62: 1387- 1396.

O'Hara, J. E. 2002. Revision of the Polideini (Tachinidae) of America north of Mexico. *Studia dipterologica.*, 10: 1-170.

O'Hara, J. E. 2003. General information about tachinid flies. Internet address: http://www.nadsdiptera.org/Tach/Gen/tachintr.htm

O'Hara, J. E. 2008. Tachinid flies (Diptera: Tachinidae). Pp. 3675–3686. In Capinera, J. L. ed. *Encyclopedia of Entomology*, 2nd Edition. Springer Netherlands, Dordrecht., 4346 pp.

O'Hara, J. E. 2010. World genera of the Tachinidae (Diptera) and their regional occurrence.Version 5. 0. PDF document, 74 pp. Available online: http://www.nadsdiptera.org/Tach/Genera/generahom.htm.

O'Hara, J. E. 2011. World genera of the Tachinidae (Diptera) and their regional occurrence.Version6.0.PDFdocument,75pp.Availablefrom:http://www.nadsdiptera.org/Tach/Genera/ Gentach_ver6.pdf (accessed 3 June 2012).

O'Hara, J. E. 2012. Review of *Euthera* (Diptera: Tachinidae) in North America with the description of a new species. *Canadian Entomologist.*, 144: 206- 215.

O'Hara, J. E. 2012. Review of the tachinid fly diversity in the Gila National Forest, New Mexico. New Mexico Issue 26, 2013 The Tachinid Times 55 Botanist, Special Issue No. 3[=Proceedings of the Third Natural History of themGila Symposium, October 14–16, 2010,Western New Mexico University, Silver City, New Mexico: 32–41.

O'Hara, J. E. 2013. History of tachinid classification (Diptera, Tachinidae). *ZooKeys.*, 316: 1–34

O'Hara, J.E. 1993. Revision of the species of *Frontiniella* Townsend (Diptera: Tachinidae). *Canadian Entomologist*. 125: 11– 45.

O'Hara, J.E. 1994. Revision of Nearctic species of *Ceromya* Robineau-Desvoidy (Diptera: Tachinidae). *Can. Ent*. 126:775-806.

Osten Sacken, C.R. 1882. Enumeration of the Diptera of the Malay Archipelago collected by Prof. Odoardo Beccari etc. Supplement. *Annali Mus. Civ. Stor. Nat. Genova.* 18:10-20. (I-II).

Pandharbale, A.R. and Sathe T.V., 2001. On a new species of the Genus *Syntomis* (Syntomidae : Lepidoptera) from the environment of Western Ghats (Satara District). *Indian J. Environ. and Ecoplan.*, 5(2), 601-602.

Perry, I. 2008. A further record of *Opesia grandis* (Egger, 1860) (Diptera, Tachinidae) from Cambridgeshire. *Dipterists Digest* (2nd Series) 15: 4.

Qi, L., Hou, P. and Zhang, C. T. 2013. The first description of the male of *Sericozenillia albipila* (Mesnil) (Diptera, Tachinidae), a newly recorded genus and species from China. *Acta Zootaxonomica Sinica,* 38: 457–460.

Raper, C.M., Smith, M.N. and Gibbs, D.J. 2006. *Carcelia laxifrons* Villeneuve (Diptera, Tachinidae) new to Britain and a revised key to the British *Carcelia* species. *Dipterists Digest* (2nd series) 13: 49–55.

Rayner R. and Raper C. 2001. Die Raupenfliegen (Diptera:Tachinidae) Mitteleuropas: Bestimmungstabellen und Angaben zur Verbreitung und Ökologie der einzelnen Arten. StuttgarterBeiträge zur Naturkunde (A) 506: 1-170. Online Authorized Version of English translation by Rayner R. and Raper C.: Tschorsnig H.-P. and Herting B. 2001: The Tachinids (Diptera: Tachinidae) of Central Europe: Identification Keys for the Species and Data on Distribution and Ecology,http://tachinidae.org.uk/site/downloads.php, 1994.

Reinhard, H. J. 1934. New North American Tachinidae. *Bull. Brooklyn ent. Soc.,* 29:186-195.

Reinhard, H. J. 1935. North American two-winged flies of the genus Doryphorophaga (Tachinidae, Diptera). *J. N.Y. Ent. Soc.,* 43: 387-394.

Richter, V. A. 2008. A new genus and a new species of tachinids (Diptera, Tachinidae) from deserts of Middle Asia. [In Russian.] Entomologicheskoe Obozrenie 87: 663–667. *English translation in Entomological Review* 88(6): 727-729.

Richter, V. A. 2009. A new genus and a new species of tachinids (Diptera, Tachinidae) from the south of Middle Asia. *Entomologicheskoe Obozrenie.,* 88: 689- 692.

Richter, V. A. 2012. A New Genus and a New Species of Tachinids (Diptera: Tachinidae) larvae. *Florida Entomologist,* 89: 497- 501.

Richter, V.A. 1992. Two new Palaearctic tachinid species of the tribe Elfiini (Diptera: Tachinidae). *Zoosystematica Rossica* 1: 145-147.

Richter, V.A. 2004. A new species of the tachinid genus ThelairaRob.-Desv. (Diptera, Tachinidae) from Tajikistan [In Russian]. *Entomologicheskoe Obozrenie* 83: 905-908.

Richter, V.A. and B. Zhumanov. 1994. The tachinid genus *Goniophthalmus* new to the fauna of Middle Asia (Diptera, Tachinidae) [In Russian]. *Zoosystematica Rossica,* 3: 146.

Robineau-Desvoidy, J. B. 1830. Essai sur les Myodaires. *Mem. Pres. Div. Sav. Acad. Sci. Inst. Fr.,* 2: 1-183.

Rondani, C. 1845. Genera italic ordinis Dipterorum ordinatim disposita et distinct et in familias et stirpes aggregate. *Dipterologiae italicae prodromus* 1,226 pp. Parma.

Roonwal, M.L., G.D. Bhasin and G.D. Pant, 1950. A systematic catalogue of the main identified entomological collection at the Forest Research Institute, Dehra Dun. Parts 1-3. *Indian Forestry,* 76(11), 498-505.

Root, R. 1973. Organization of a plant–arthropod association in simple and diverse habitats, the fauna of collards (*Brassica oleracea*). *Ecological Monographs.*, 43: 95–124.

Sabrosky, C. W. and Araud, P.H. 1936. Manual of Myiology. *Itaquaecetuba,* Sao Paulo. Part-III- IV. 255-309pp.

Sabrosky, C. W. and Araud, P.H. 1938. Manual of Myiology. *Itaquaecetuba,* Sao Paulo. Part-VII. 434pp.

Sabrosky, C. W. and Araud, P.H. 1965. Family Tachinidae (Larvaoridae). *In* stone et. al. A catalog of the Diptera of America north of Mexico. *Agric. Handb.* No. 276, 1696pp.

Samways, M. J. 2005. Insect Diversity Conservation. Cambridge University Press, Cambridge, UK, 342 p.

Sathe T. V. 2007. Biodiversity of Wild Silkmoths From Western Maharashtra, India. *Bull. Ind. Acad. Seri.,* 2(1), 21-24.

Sathe T. V. 2008. Mass production Technique for *Campoletis chlorideae* (Uchida). *Biotechnological Approaches in Entomology,* 3, 64-74.

Sathe T. V. 2012. Biodiversity of Ichneumonid flies (Hymenoptera: Ichneumonidae) from Western Ghats, Maharashtra. *The scientific Temper,* 3 (1 & 2), 47-50.

Sathe T. V. 2012. Pests of Ornamental Plants. Daya publishing house, New Delhi. Pp. 1-199.

Sathe T. V. 2013. Dragon flies production technology. Daya publishing house, New Delhi. Pp.1-102.

Sathe T. V. 2014. Ecology, epidemiology and control of Sand flies from Kolhapur region, India.

Sathe T.V. 2009. A textbook of forest Entomology. Daya publishing house, New Delhi. Pp. 1-234.

Sathe T.V. 2010. Biocontrol approaches in mulberry pest management. *Application of Biotechnology in Sericulture,* 21, 229-241.

Sathe T.V. 2010. Biodiversity of Damselflies (Odonata) from Koyna Dam and around area. *Flora & Fauna,* 16, 68-72.

Sathe T.V.2011. Ecology of Mosquitoes from Kolhapur district, India. *International Journal of Pharma & Biosciences,* 2 (4) (B), 103-111.

Sathe, T. V.2012. Biodiversity of tachinid flies. (Diptera: Tachinidae) from western Maharashtra. *International Journal of Plant Protection,* 5, 368-370.

Sathe, T.V. 1992. Natural enemies of some insect pests of economic importance. *Oikoassay,* 9, 15-17.

Sathe, T.V. 2003. Biodiversity of braconid pest biocontrol agents from Western Maharashtra. *Bull. Bio. Sci.,* 1, 73-75.

Sehnal, P. 1998. A new species of *Borgmeiermyia* Townsend, 1935, from Paraguay (Insecta: Diptera: Tachinidae). Peter Sehnal, Naturhistorisches Museum Wien, 2. Zoologische Abteilung, Burgring 7, A-1014 Vienna, Austria, 349-353.

Shima H, Chao CM, Zhiang WX. 1992. The genus *Winthemia* (Diptera, Tachinidae) from Yunnan Province, China. *Japanese Journal of Entomology*, 60(1): 207-228.

Shima, H. 1983. A new species of *Oxyphyllomyia* (Diptera, Tachinidae) from Nepal, with reference to the phylogenetic position of the genus. *Annot. zool. jpn.*, 56: 338-350.

Shima, H. 1987. A Revision of the Genus *Isosturmia* Townsend (Diptera, Tachinidae) *Bull. Kitakyushu Mus. Nat. Hist.*, 6: 213-237.

Shima, H. 1996. A systematic study of the genus *Cavillatrix* Richter (Diptera, Tachinidae). *Bulletin of the Graduate School of Social and Cultural Studies, Kyushu University.*, 2: 133-148.

Shima, H. and Chao, C.-m. 1988. A new genus and six new species of the tribe Goniini (Diptera: Tachinidae) from China, Thailand and New Guinea. *Systematic Entomology*, 13: 347–359.

Shima, H. and Tachi, T. 2008. New species of the genus ParavibrissinaShima (Diptera: Tachinidae) from Southeast Asia and South Pacific. Zootaxa, 1870: 43–60.

Shima, H. and Tachi, T. 2009. Description of a new species of the genus *Setalunula* Chao and Yang (Diptera, Tachinidae) from Japan. *Bulletin of the National Science Museum*. Series A (Zoology), 35: 233–242.

Shima, H., Han, H. Y. and Tachi, T. 2010. Description of a new genus and six new species of Tachinidae (Diptera) from Asia and New Guinea. *Zootaxa.*, 2516: 49-67.

Shivaramkrishnan, K.G., Venkataraman, K., Morthy R.K. Utkash, G. and K.A. Subramanian, 2000. Aquatic insect diversity and ubiquity in the streams of the Western Ghats, India. *J. Indian Inst. Sci.*, 80, 537-552.

Simpson, M. G. 2010. "Chapter 1 Plant Systematics: an Overview". *Plant Systematics* (2nd ed.) Academic Press. ISBN 978-0-12-374380-0.

Skevington, J. H. and Dang, P.T. 2002. Exploring the diversity of flies (Diptera). *Biodiversity.*, 3(4): 3-27.

Smit, J.T. and Zeegers, T. 2002. The Tachinidae and Oestridae (Diptera) of Madeira, with description of a new species. Stutt. Beitr. *Naturk.* (A): 642: 1-11.

Son, J. K., Do, N. X. and Park, C. G. 2008. Seasonal parasitism of *Riptortus clavatus* Thunberg (Heteroptera: Alydidae) by *Dionaea magnifrons* (Herting) (Diptera: Tachinidae). *Journal of Asia-Pacific Entomology.*, 11: 191–194.

Stireman J.O., O'Hara J. E., and Wood D. M. 2006. Tachinidae: Evolution, Behavior, and Ecology. *Annu. Rev. Entomol.*, 51(1): 525-555.

Stireman, *et al.*, 2006. Tachinidae: Evolution, Behavior, and Ecology. *Annu. Rev. Entomol.*, 51: 525–55.

Stireman, J. O. 2002. Host location and selection cues in a generalist tachinid parasitoid. *Entomologia Experimentalis et Applicata.*, 103: 23–34, Kluwer Academic Publishers. *Printed in the Netherlands.*

Sun, X. 1993. Notes on two new species of the genus *Calozenillia* Townsend from China (Diptera: Tachinidae) [In Chinese]. *Sinozoologia,* 1993: 441-443.

Sun, X. and Marshall, S. A. 1995. Two new species of *Cylindromyia* Meigen (Diptera, Tachinidae), with a review of the eastern Palaearctic species of the genus. *Studia dipterologica,* 2(2): 189–202.

Sun, X.-k. and Chao, C.-m. 1992. A new species of *Zenilliana* from China (Diptera: Tachinidae) [In Chinese with English summary]. *Sinozoologia,* 9: 331–333.

Sun, X.-k. and Chao, C.-m. 1993. Notes on the genus *Pexopsis* Brauer and Bergenstamm from China (Diptera: Tachinidae) [In Chinese with English summary]. *Sinozoologia,* 10: 445–458.

Sun, X.-k. and Chao, C.-m. 1994. A new genus and species of the tribe Sturmiini from China (Diptera: Tachinidae) [In Chinese with English summary]. *Acta Zootaxonomica Sinica,* 19: 480–483.

Systematics and molecular phylogeny of Mosquitoes from Western Maharashtra. Phd Thesis. Shivaji University, Kolhapur, India.

Tachi, T. 2013. A new species of the genus *Trichoformoso-myia* Townsend from Malaysia (Diptera: Tachinidae). *Zootaxa,* 3702: 61–70.

Tachi, T. and Shima, H. 2006. Review of genera *Entomophaga* and Proceromyia (Diptera: Tachinidae). *Annals of the Entomological Society of America* 99: 41-57.

Tanaka, C., Kainoh, Y., and Honda, H. 1999. Comparison of oviposition on host larvae and rubber tubes by *Exorista japonica* Townsend (Diptera: Tachinidae). *Biol. Control,* 14: 7-10.

Tanaka, C., Kainoh, Y., and Honda, H. 1999. Physical factors in host selection of the parasitoid y *Exorista japonica* Townsend (Diptera: Tachinidae). *Appl. Entomol. Zool.,* 34: 91–97.

Thompson, F. C. 2005. Biosystematic Database of World Diptera. Version 7.5, http://www.diptera.org/biosys. htm.

Toma, R. 2008. A new species of *Leschenaultia* Robineau-Desvoidy (Diptera, Tachinidae) from Venezuela and new geographical records. *Revista Brasileira de Entomologia,* 52(3): 353-354.

Toma, R. 2008. A new species of *Leschenaultia Robineau-Desvoidy* (Diptera, Tachinidae) from Venezuela and new geological records. *Revista Brasileira de Entomologia,* 52(3): 353-354.

Tothill, J. D. 1924a. A revision of the Nearctic species of the genus Gonia(Diptera, Tachinidae). *Can. Ent.,* 56: 196-200, 206-212.

Tothill, J. D. 1924b. A revision of the Nearctic species in the genus Fabriciella(Tachinidae). *Can. Ent.,* 56: 257-269.

Tschorsnig, H. P. and Herting, B. 1994. The Tachinids (Diptera: Tachinidae) of Central Europe: Identification Keys for the Species and Data on Distribution and Ecology. http://tachinidae.org.uk/site/downloads.

Tschorsnig, H.P. 1993. A new species of the genus *Bithia* Robineau-Desvoidy (Diptera: Tachinidae) from Tadjikistan and Kazakhstan. *Stutt. Beitr. Naturk.* (A) 494, 4 pp.

Tunca, H., Kara, K. and Ozkan, C. 2009. Two Tachinid (Diptera: Tachinidae) parasitoids of *Diprion pini* (L.) (Hymenoptera: Diprionidae), along with a new record for the Turkish Tachinidae fauna. *Turk. J. Zool.*, 33: 1-4.

Van den Bosch, R., Messenger, P. S., and Gutierrez, A. P. 1982. An introduction to biological control. Plenum Press, New York and London. 247 pp.

Van Embden 1960. Keys to Ethiopian Tachinidae-III Macquartiinae. *Proc. Zool. Soc. Lond.* 134:313-487.

Verbeke, J. 1963. The structure of the male genitalia in Tachinidae (Diptera) and their taxonomic value. *Stuttgarter Beiträge zur Naturkunde,* 114: 1–5.

Vincent, L. C. 1985. "The first record of a tachinid fly as an internal parasitoid of a spider (Diptera: Tachinidae) Araneae; Anthrodiaeti-dae". *Pan-pacific Entomology.,* 61: 224-225.

Wang, Q. and Zhang, C.T. 2012. A new species of the genus *Neaera* Robineau-Desvoidy (Diptera, Tachinidae) from Sichuan, China. *Acta Zootaxonomica Sinica.,* 37: 829–833.

Wang, Q., Fan, H. and Zhang, C. T. 2012. Notes on the genus *Cyrtophleba* (Diptera: Tachinidae) of China. *Entomo taxonomia.,* 34: 356–360.

Webber, R. T. 1941. The synopsis of the tachinid flies from genus *Tachinomyia* and description of new species. *Proc. U. S. Natn. Mus.*, 90(3108): 287-304.

Weseloh, R. M. 1980. Host recognition behavior of the Tachinid parasitoid, *Compsilura concinnata. Ann. Ent. Soc. Am.,* 73: 593–601.

Williams, S. C., Arnaud, P. H. and Lowe, G. 1990. "Parasitism of Anuroctonus phaiodactylus (Scorpiones: Vaejovidae) by Spilochaet- osoma californium Smith (Diptera: Tachinidae) a review of parasitism in scorpios". *Myia* 5: 11-27.

Wood, D. M. 1985. A taxonomic conspectus of the Blondeliini of North and Central Americaand the West Indies (Diptera: Tachinidae). *Mem. Ent. Soc. Can.,* 132: 1-130.

Wood, D. M. 1987. "Tachinidae". In: *Manual of Nearctic Diptera,* 2. Pp. 1193- 1269. (eds. J.E.B.V. McAlpine, G.E. Peterson, H.J).

Woodley *et al.,* 2008. *Lobomyia neotropica,* a new genus and species of Tachinidae (Diptera) from the Neotropical Region. *Zootaxa,* 1783: 31–39.

Woodley, N. E. 1994. A new species of Lydella (Diptera: Tachinidae) from Mexico with a discussion of the definition of the genus. *Bulletin of Entomological Research,* 84: 131-136.

Woodley, N. E., Paul, H. and Arnaud, J. R. 2008. Lobomyia neotropica, a new genus and species of Tachinidae (Diptera) from the Neotropical Region. *Zootaxa,* 1783: 31–39.

Xu, S.h. and Lou, J.x. 2000. Notes on two genera of Microteryini new to China descriptions of with two new species (Hymenoptera: Encyrtidae) [In Chinese]. *Acta Zootaxonomica Sinica,* 25: 199-203.

Yang, L.-l. and Chao, C.-m. 1990. Four new species of tribe Blondeliini from the Nanling Mountains of China (Diptera: Tachinidae). [In Chinese with English summary]. *Sino-zoologia,* 7: 307–313.

Yao *et al.,* 2010. Notes on species of the tribe Exoristini from China (Diptera: Tachinidae), *Journal of Shenyang Normal University (Natural Science),* 28(4):530-533.

Yeates, D. K. and Wiegmann, B. M. 1999. Congruence and Controversy: Toward a Higher Level Classification of Diptera. *Annual Review of Entomology,* 44: 397–428.

Zeegers Thao and Bahareh Majnon Jahromi 2015. First record of the genus *Blepharella* (Diptera: Tachinidae) from the western Palaearctic region

Zhang, C.-t. and Fu, C. 2012. Three new species of Dinera Robineau-Desvoidy from China (Diptera: Tachinidae). *Zootaxa,* 3275: 20–28.

Zhang, C.t. and Shima, H. 2004. Taxonomic study of *Dexia* Meigen from China (Diptera: Tachinidae). *Journal of Shenyang Normal University (Natural Science),* 22: 49-56.

Zhang, C.-t. and Shima, H. 2006. A systematic study of the genus DineraRobineau-Desvoidy from the Palaearctic and Oriental regions (Diptera: Tachinidae). *Zootaxa,* 1243: 1–60.

Zhang, C.-t., Wang, M.-f. and Ge, Z.-p. 2007. First record of the genus *Atylomyiain* China with two new species (Diptera,Tachinidae). *Acta Zootaxonomica Sinica,* 32: 585–589.

Zhang, C.-t., Wang, M.-f. and Liu, J.-y. 2006. A new species and a new record of the genus *Leptothelaira* from China (Diptera,Tachinidae). *Acta Zootaxonomica Sinica,* 31: 430–433.

Zhao, Z., Zhang, C.-t. and Chen, X.-l. 2012. A new species of the genus Kuwanimyia Townsend (Diptera: Tachinidae) from Zhanjiang, China. *Zootaxa,* 3238: 57–63.

Ziegler, J. 2010. Revision of the genus *Germaria* Robineau-Desvoidy (Diptera, Tachinidae) from Greece, with descriptions of two new species. *Deutsche Entomologische Zeitschrift,* 57: 43–57.

Ziegler, J., Shima H. 1996. Tachinid flies of ussuri area (Diptera:Tachinidae). *Beitrage zur Entomologie,* 46: 378-379.

Zoological Survey of India, 1983. Threatened Animals of India, Calcutta: Zoological Survey of India. pp. 307.

Index

D

E

F

G

H

L

M

N

O

P

www.ingramcontent.com/pod-product-compliance
Ingram Content Group UK Ltd.
Pitfield, Milton Keynes, MK11 3LW, UK
UKHW021952270726
14060UKWH00002B/470